Stay Curious and Keep Exploring

Stay Curious and Keep Exploring

50 Amazing, Bubbly, and Creative Science Experiments to Do with the Whole Family

EMILY CALANDRELLI

CHRONICLE PRISM

Library of Congress Cataloging-in-Publication Data available.

ISBN 978-1-7972-1622-5

Manufactured in China.

Design by Erin Jang.
Illustration by Cachetejack.
Typesetting by Sandy Lynn Davis. Typeset in Avenir, Belizio, Futura Round, GT Maru, GT Ultra, GT Walsheim, Gordon, Too Much Opaque, Gooper, Juniper, Poplar, and Village.

10 9 8 7 6 5 4 3 2 1

Chronicle books and gifts are available at special quantity discounts to corporations, professional associations, literacy programs, and other organizations. For details and discount information, please contact our premiums department at corporatesales@chroniclebooks.com or at 1-800-759-0190.

CHRONICLE PRISM

Chronicle Prism is an imprint of Chronicle Books LLC, 680 Second Street, San Francisco, California 94107

chronicleprism.com

For the kids who dream of becoming scientists

Table of Contents

Introduction

Hello, little scientists! Emily here. Welcome to *Stay Curious and Keep Exploring*, my book of experiments that are fun for the whole family. This book contains 50 of my favorite science experiments that you can all do together, usually with materials you already have at home!

Some of these experiments will POP, some will bubble over, some will GLOW in the dark, and some may be a little spooky! For each experiment, I'll tell you exactly what you need, and my little scientists and I will walk you through it step by step. You may need an official Lab Assistant (an adult) for some of them, so make sure you have one handy. Any experiment that requires adult supervision will say "Lab Assistant (an Adult!) Required" at the top.

As you flip through the pages, you'll find that this book is filled with more than just science experiments. You'll learn answers to questions like, "Why is ketchup so hard to get out of the bottle?" and "Why do my ears pop sometimes?" You'll read about important women in science who changed the course of history, and you'll be asked to create your own hypotheses about the experiments—what do YOU think will happen? Look out for the **words highlighted in yellow** on each page. These are my favorite science words that I've included in the Science Glossary in the back of the book. And, of course, I'll share some fun stories from my time on *Emily's Wonder Lab!*

I want this book to encourage you to stay curious about the world around you. I want you to ask questions, make hypotheses, and test your ideas in the real world. Keep exploring anything and everything around you. Science is so fun and exciting because it's all about learning how the world works. What may seem like magic is actually science. And YOU can learn it all! This book will help you become a little scientist yourself. So, what do you say? Should we dive into the experiments? Let's go!

9

Space Slime

Since meteorites fall to Earth all the time, how do I know if a rock in my backyard is from outer space?

ABOUT THE EXPERIMENT

Meteorites do fall to Earth, but it is rare to find them in your backyard. However, if you do find a rock that you think might be a meteorite, you should grab a magnet to see if it's attracted to it. Many meteorites come from asteroids, and many asteroids have a high percentage of iron in them. Iron in meteorites is attracted to magnets. So, if the magnet test works, you should contact your local museum and see if they might be able to verify whether or not you found a rock . . . from space!

This experiment will show you how to make your very own slime . . . from space! Well, it won't *actually* be from space, but it will be attracted to a magnet just like meteorites—and we can pretend, right?

Materials
↓

⅓ cup (80 ml) white glue

2 tablespoons water

¼ teaspoon baking soda

Bowl

Spoon

2 tablespoons iron oxide powder
(you can buy this online)

¾ tablespoon eye contact solution
(must contain boric acid and sodium borate)

A strong magnet

INSTRUCTIONS

1. Pour the glue, water, and baking soda into your bowl and mix them together with a spoon.

2. Use your spoon to mix in the iron oxide powder.

3. Add the contact solution and stir again.

4. Take your magnet and bring it close to your slime. Try to get within 1 inch (~2 cm), and then ½ inch (~1 cm), and see if you can make your slime move to follow the magnet. Be careful not to get the slime on the magnet. It's not a big deal if you do, but it is a little hard to get off.

As you bring your magnet really close to your slime, you should see small parts of the slime move toward the magnet (not a lot of the slime, but a small amount right on the surface).

THE SCIENCE

 Whoa, this thing looks like an alien! It's aliveeeeee!

 It certainly looks that way, doesn't it?

 What is making it move around?

 The iron oxide we added to the slime is the secret ingredient. Iron is attracted to magnets. This is because when iron is in the presence of a magnetic field, the atoms start to align their electrons in the direction of that magnetic field—this temporarily makes the iron a magnet too. Then those two magnets (your iron and your other magnet) become attracted to each other.

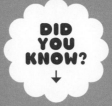
The best way to hunt for meteorites is looking in places where a dark piece of rock would really stand out. If you go to an icy place like Antarctica or the middle of a light-colored, sandy desert and you see a dark, black rock just sitting on top of the land, chances are that rock came from space!

Most meteorites found on Earth come from the asteroid belt, but sometimes you can find one that came all the way from the Moon or Mars. How did a rock from the Moon or Mars get here? Well, a big asteroid had to slam into the Moon or Mars, kick up a bunch of rocks, and send them flying into the solar system. Some of that debris landed right here on Earth, although those types of meteorites are much more rare.

What's the difference between a meteoroid, a meteor, and a meteorite? A meteoroid is a rock flying through space, a meteor is a space rock flying through Earth's atmosphere, and a meteorite is a space rock that's landed on the ground.

Fizzy Rocket

What makes a rocket move upward?

ABOUT THE EXPERIMENT

A rocket moves upward because it creates an explosion downward. That's **Newton's Third Law of Motion** : For every action, there's an equal and opposite reaction. This is the same reason why blowing up a balloon and letting it go (before tying it off) will make the balloon fly in the opposite direction the air moves in. With this experiment, we'll see Newton's Third Law in action. It is so exciting, and I scream every time a film canister launches! You can certainly do this one indoors, but it might be more fun to launch them outside. The canister will get about a tablespoon of water and fizz wherever it launches, so make sure it's on something you can easily wipe off.

Materials
↓

Goggles or protective eyewear
(to protect your eyes when launching a film canister)

Film canister rocket
(these can be purchased online; search for this term)

1 spoonful water

Antacid tablets

MAKE A HYPOTHESIS
→ **What would happen if you added less water in your film canister?**
→ **What about more water?** →

INSTRUCTIONS

1. Put on your goggles.
2. Open your film canister and add the water.
3. Place one antacid tablet inside.
4. Quickly place the lid on and then flip your canister over on a flat surface so that the lid is facing downward.

When you flip over your closed film canister, you will need to wait for 5 to 15 seconds and then all of a sudden—POP! The larger part of your cannister will shoot straight up into the air. It may even hit the ceiling!

Warning:
Be sure to wear safety goggles and don't stand directly over your film canister because it will launch straight upward!

→

Make your own guess, then flip upside down to read!

When you add less water, the canister will take longer to pop (because there's more volume for bubbles to fill), but it will launch higher than before. This is because there's now more air in the container shooting downward. When you add more water, your canister will pop faster but there will be less air moving downward (and more friction between the canister and the water), so your rocket won't launch as high.

THE SCIENCE

 That went so high! What made that launch?

 When you add an antacid tablet to water, it starts to fizz and bubble. All of those bubbles build and build and increase the air **pressure** inside your canister. At some point the pressure becomes so great that the lid simply can't hold on anymore and your canister shoots upward. For every action—the rush of air DOWNward—there's an equal and opposite reaction—the canister moves UPward!

 Cool! But why does water make antacid fizz?

 It's a chemical reaction! Inside antacid is baking soda and citric acid. Baking soda is a base, and citric acid, which is the same stuff that's in lemon juice, is an acid. But of course! It's in the name! Check out the pH scale in the back of the book to learn about other acids and bases! Without water, the tablet doesn't fizz because all of the molecules are mostly standing still and not interacting. But as soon as you add water, they start partying together and it kick-starts an acid-base **chemical reaction** that creates a lot of carbon dioxide bubbles. Perfect for launching rockets! Well, small ones, anyway.

NOTES!

_____

_____

_____

_____

_____

_____

WHO TO KNOW
↓

Mae Jemison was the first African American woman to travel into space. She is an engineer, a physician, and an astronaut who flew on a space shuttle mission that orbited the Earth for nearly eight days in 1992 and had a guest appearance on the popular TV show *Star Trek: The Next Generation* the following year!

Balloon Rocket

Lab Assistant (an Adult!) Required

How does jumping work?

ABOUT THE EXPERIMENT

I want you to try something right now. I want you to jump as high as you can in the air.

What did you do to make that happen? You probably bent your legs and pushed down on the ground really hard. You are practicing **Newton's Third Law of Motion** : For every action there's an equal and opposite reaction. Imagine you're standing on a skateboard. If you throw your heavy school bag, you roll in the opposite direction! The heavier the bag, or the faster you throw it, the further you roll. By pushing against the bag in one direction, you are forcing the skateboard (and you!) to roll in the other direction!

With this experiment, you'll be able to put this law of motion into action and create your own balloon rocket. If you love space like I do, you'll love this one!

Materials
↓

Tape

**10- to 15-foot
(3- to 4.5-m)
length of string**
(any string will do but it
has to be thin enough to go
through a straw)

Straw

Balloon

→

INSTRUCTIONS

1. Have your lab assistant (an adult!) tape one end of your string to something high up (this could be your ceiling, or a railing on your stairs, or simply the top of a chair).

2. Bring the other end of the string far away so that the string remains straight.

3. Put that end of the string through a straw.

4. Blow your balloon all the way up.

5. While pinching the end, place your balloon next to your straw.

6. While you hold the balloon, have your lab assistant tape the straw onto your balloon with two small strips of tape.

7. Bring your string down low to the ground, make sure that it is taut (pulled tight), and then let go of your balloon.

As you let go of your balloon, the air should fly out, zipping your balloon rocket along the string.

Valentina Tereshkova was the first woman in space. She was only twenty-six years old when she orbited the Earth forty-eight times over the course of three days.

TRY
THIS
↓

Get creative by adding a payload (tape on a cotton ball for an astronaut) or adding multiple boosters (tape together more balloons).

Or we can turn this into a game. Tape two different strings to the ceiling (or you can try one that is level this time) and then set up two different straws and balloons. Have a race to see which balloon rocket can get to the end first.

THE SCIENCE

 Wow, it followed the string to the ceiling!

 That's Newton's Third Law of Motion happening right in front of you! For every action—in this case, air flying down out of your balloon—there's an equal and opposite reaction—the balloon moving upward in the opposite direction of the air.

 Is that how a real rocket works?

 Kind of. This balloon is just using air as a propellant, so it can't go very high. A rocket will light different chemicals on fire, creating a huge explosion downward, which pushes the rocket upward. They both use propellants, just different types.

NOTES!

_____

_____

_____

_____

_____

_____

_____

Alien Hovercraft

Lab Assistant (an Adult!) Required

How does an air hockey table work?

ABOUT THE EXPERIMENT

The next time you play a game of air hockey, look at the playing field. Do you see a bunch of holes? Well, when you turn the game on, you'll hear a loud noise—that's the fan turning on. Put your hand over those little holes. Do you feel it? That's air. The air is the secret to this game. A hockey puck will glide on top of that air, just like an alien hovercraft! In this experiment, we're going to recreate this same effect with a DVD and a balloon. When I did this on *Emily's Wonder Lab*, I was surprised by how quickly these can move. Mine flew right off the edge of the table! Make it your own by decorating your hovercraft with permanent markers.

Materials
↓

Sports bottle cap
(the kind that you pull to open and push to close)

Hot glue gun and glue

CD or DVD
(you can buy these online if you don't have them around)

Balloon

MAKE A HYPOTHESIS → **What would happen if you put your hovercraft on different surfaces?** → **Try carpet, a wooden table, or a glass table.**

INSTRUCTIONS

1. Push the top of your bottle cap in so that it's in the closed position.
2. Have your lab assistant (an adult!) use a hot glue gun to glue the bottle cap to the center of your DVD. Let it dry.
3. Blow up your balloon.
4. While pinching the opening to stop air from escaping, carefully place the balloon over your closed bottle cap.
5. Set your hovercraft on a flat surface, like a hard floor or a smooth table.
6. Pull the bottle cap open and push your hover-craft forward.

As soon as you pull your cap up, you should feel your hovercraft slightly move. If you tap it slightly, it should glide across your surface. If it doesn't work, make sure your cap is completely open and the surface you are on is completely flat.

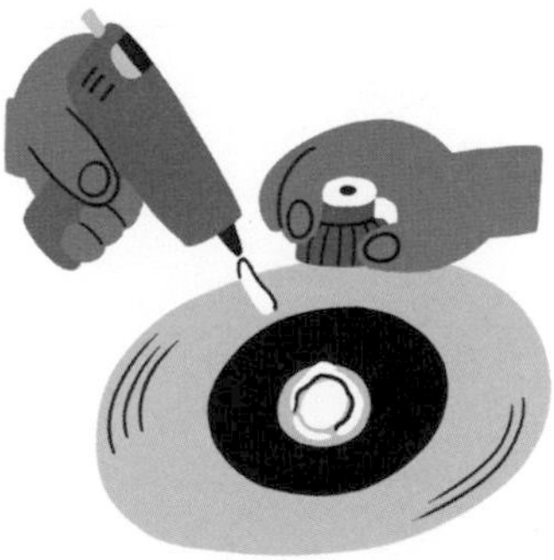

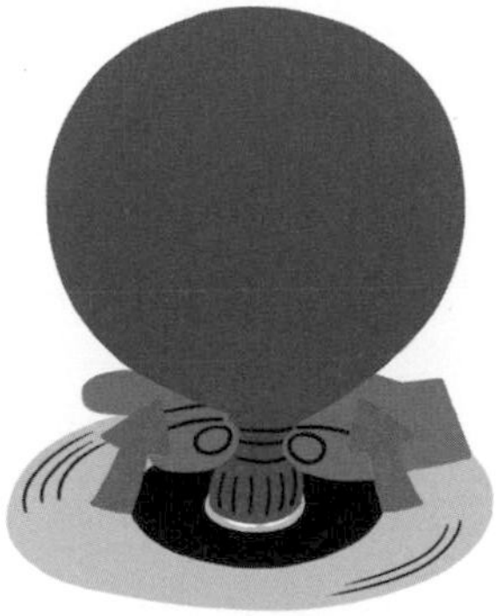

THE SCIENCE

Look at it hover! How is it floating like that?

Ah, it's not actually floating. When the DVD slides against the table, there is a lot of friction. This stops the DVD from sliding far. The balloon is pushing air out to the bottom of the DVD, creating a cushion of air beneath your hovercraft and lifting it up. That air will provide less friction (this means there is less resistance between the ground and your hovercraft) and allow your DVD to slip and slide. Have you ever walked on an icy sidewalk? It's slippery, isn't it? That's because ice provides less friction on your shoes than the dry ground. This is very similar to that—air is just a bit slippery.

But there's air all around us. Why can't it glide forever?

Well, it's the air leaving your balloon that's forcing your hovercraft to hover! Your hovercraft can gently glide away as long as the balloon still has air to push out. As soon as your balloon runs out of air, your DVD will fall down to the ground and won't be able to move anymore because whatever surface you're on has too much friction.

→ (Make your own guess, then flip upside down to read!)

*The smoother the surface, the better your hovercraft will work. Remember, this is all about **friction**. Something that is smoother will create less friction between the air and the surface itself, allowing the hovercraft to smoothly glide along.*

SPOOKY SCIENCE!

Barfing Pumpkin

How do kids make those erupting volcanoes at science fairs?

ABOUT THE EXPERIMENT

This is a great one to do outside and is especially perfect right after Halloween with a jack-o'-lantern that may be just starting to rot, but you don't want to throw it out just yet. This is also the same science experiment that people use to create those erupting volcanoes you may have seen at science fairs—instead of putting the ingredients in a pumpkin, they just put them inside a homemade volcano. Grab a jack-o'-lantern and head outside—we're going to make a pumpkin barf!

I'M NOT FEELING SO GOOD

Materials
↓

Jack-o'-lantern

2 cups (460 g) baking soda

Large glass

2 cups (480 ml) vinegar

3 squirts of dish soap

5 drops of green food coloring
(optional)

MAKE A HYPOTHESIS → What do you think would happen if you didn't add the soap? → Try it out! →

INSTRUCTIONS

1. Bring your jack-o'-lantern outside.
2. Place the baking soda inside the jack-o'-lantern.
3. In a large glass, mix the vinegar, dish soap, and food coloring (if using).
4. Slowly pour the contents of the glass inside the jack-o'-lantern.

You should see your ingredients immediately bubble over, creating green frothy bubbly barf that quickly falls out of your jack-o'-lantern's mouth!

THE SCIENCE

 Awesome! My pumpkin barfed green foamy liquid all over the place. How did that happen?

 Baking soda is a base and vinegar is an acid. When you mix these two ingredients together, you start a chemical reaction that creates carbon dioxide bubbles. (For more examples of acids and bases, check out the pH scale in the back of the book!)

 What does the soap do?

 The soap traps a lot of that carbon dioxide gas into soapy bubbles, which makes the barf a little more . . . frothy!

 Gross!

NOTES!

_____

_____

_____

_____

_____

_____

(Make your own guess, then flip upside down to read!)

The barf looks a little different, doesn't it? It's a little less frothy. That's because we no longer have the soap to trap all that air. It still looks bubbly, just not as thick as before. Some of that carbon dioxide gas is simply releasing into the air instead of being trapped by the soap.

Flying Tea Bag Ghosts

Lab Assistant (an Adult!) Required

How do hot air balloons work?

ABOUT THE EXPERIMENT

My little scientists absolutely loved this experiment on *Emily's Wonder Lab*. At first, it doesn't seem like anything is happening, and then BOOM! Your flying tea bag ghosts fly to the ceiling! It's incredibly easy to do (with adult supervision, of course!) and very fun to watch. This one only takes a few minutes to prepare, and the cleanup is quick and easy. The reason it works is similar to the reason a hot air balloon floats. When you heat air up, it becomes less **dense** (the molecules want to spread out more). Because the hot air is less dense, it rises among the cooler air around it. That's why you see hot air balloon pilots lighting the air inside the balloon with fire—they need to keep the air inside the balloon at a relatively low density, so they can rise into the sky.

THE SCIENCE

  Why did it rise to the ceiling?

  What's that stuff falling back down?

It rises to the ceiling because your tea bag is like a hot air balloon. When the tea bag was lit on fire, it started heating the air inside the tea bag and all around it. That hot air is rising in a room—even if you can't see it. The tea bag rises along with it because the tea bag itself is made up of such light material that it becomes easily carried by the hot air.

You'll notice that part of the tea bag will fall back down from the ceiling. If you catch it in your hands you can feel how light it is. It will also stain your hands black! That's because what you're holding is ash—carbon from the tea bag that was left over after the burning happened.

INSTRUCTIONS

1. Using scissors, cut off the very top of the tea bag (where the string is connected).
2. Empty the contents of the tea bag and discard.
3. Squeeze the sides of the bag and turn your tea bag into a cylinder.
4. Use markers to draw a spooky ghost on your tea bag.
5. Set your tea bag on a flat surface (you may need to use your scissors to even out the bottom of your tea bag if it keeps falling over).
6. Have an adult use a lighter to light the top of the tea bag, and then stand back!

You should see the tea bag quickly light on fire and burn from the top to the bottom. Right before it reaches the bottom, the tea bag will float up, up, up to the ceiling!

WHO TO KNOW ↓

Sally Ride was the first American woman to travel to space. She flew on NASA's space shuttle in 1983 and used a robotic arm to place satellites in orbit around the Earth.

Disappearing Act

How does nail polish remover work?

Lab Assistant (an Adult!) Required

ABOUT THE EXPERIMENT

N ail polish remover has something called acetone in it. This is what gives nail polish remover that pungent smell! Acetone is actually one of the main ingredients in nail polish too, but nail polish also has other **chemicals** in it that cause it to harden after it dries. After you apply the nail polish, the acetone quickly evaporates away, allowing the other ingredients in there to harden that beautiful new color onto your nails. When you put nail polish remover on a cotton ball and apply it to your fingernail, the acetone wiggles its way back into the nail polish and turns the nail polish back into its liquid state. Basically, you're dissolving the nail polish in acetone.

Acetone is something known as a **solvent** because it can dissolve other substances like paint and certain plastics. In this experiment, we're going to use it to dissolve . . . a ghost!

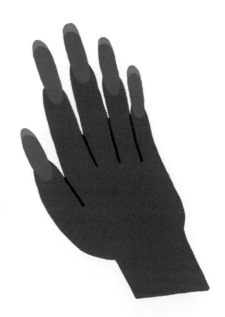

Pen

Thin Styrofoam

(this can be from a cup or packaging)

Scissors

Safety glasses

Glass or bowl

100% acetone

INSTRUCTIONS

1. With a pen, draw a ghost on your piece of Styrofoam. (I've made mine as tall as my hand. Just make sure that it will be able to fit inside your glass or bowl.)

2. With scissors, cut out your ghost.

3. Put on your safety glasses.

4. Fill your glass or bowl with 100% acetone.

5. Slowly push your ghost into the bowl or glass, submerging it under the acetone.

As you push the Styrofoam into the acetone, you should see it slowly dissolve into a gooey puddle. This happens pretty quickly. If it's not working, make sure your Styrofoam is thin enough (like the kind used in Styrofoam cups).

TRY THIS

↓

How many Styrofoam cups can you dissolve in a small bowl of acetone? Two? Five? Fifty? I've seen someone dissolve 100 Styrofoam cups in a single bowl of acetone!

Safety note:
Only adults should handle 100% acetone.

→

WHO
TO
KNOW
↓

Madam C. J. Walker was the first Black millionaire in the United States and made her fortune thanks to her understanding of the chemistry in hair care products as well as the hair care needs of other African American women. She began selling products that would help African American women suffering from hair loss and ultimately became a pioneer in the modern cosmetics industry.

THE SCIENCE

 What is happening?! My ghost is . . . melting?

 It's dissolving! This is similar to how sugar dissolves in water. The Styrofoam is simply dissolving in the acetone.

 But what is this goo that's left in my bowl?

 That's still your Styrofoam. Styrofoam has a bunch of pockets of air in it—this is what makes it so light—but when you remove those pockets of air, it condenses to be really, really small—that's the goo you see.

NOTES!

_____

_____

_____

_____

_____

Exploding Witch's Brew

Why does soda fizz when you open it?

ABOUT THE EXPERIMENT

Next time you open a can of soda, listen to that *fizz*. Do you know what that is? That sound is actually carbon dioxide bubbles rising to the surface. Soda contains dissolved carbon dioxide. The carbon dioxide wants to be in the form of a gas, but while it is sealed inside the can, the soda is under **pressure** and the carbon dioxide is forced to stay dissolved in the liquid. As soon as you open your soda can, you release this pressure, causing a lot of the carbon dioxide to turn into bubbles that rise to the surface and FIZZ.

Today, we're going to make our own carbon dioxide bubbles . . . but in a different way. This is one of my favorite experiments. You'll find this one on the spooky episode of *Emily's Wonder Lab*. We launched ten of them at once and it was one of the messiest experiments we did on the show. Oh, did I mention this is a good one to do outside? And grab your lab assistant (an adult!)—you're going to need their help with this one. →

Materials
↓

**1 to 2 cups
(230 to 460 g)
baking soda**

**Empty metal
paint can**

(you can buy these
online or sometimes your
local home improvement
store will have them)

**1 cup (240 ml)
vinegar**

Plastic cup

Safety goggles

Rubber mallet

INSTRUCTIONS

1. Pour the baking soda inside your paint can.
2. Pour the vinegar into your plastic cup.
3. Gently place your plastic cup inside your paint can, nestled into the baking soda so that it doesn't tip over.

4. Put on your safety goggles.
5. Have an adult use the mallet to hammer on the lid of the paint can (tip—hammer in a star pattern and be sure to get the lid on good and tight!).
6. Have everyone take 10 steps back.

THE SCIENCE

 That was so exciting!

 This is why it's one of my favorite experiments! What did you notice?

 Well, first I didn't see anything. And then I noticed the paint can bulging out. Why was that?

 When you flipped the paint can over, the vinegar spilled out of the cup and mixed in with your baking soda. When the acid (vinegar) mixed with the base (baking soda), it kick-started an acid-base **chemical reaction** that created carbon dioxide bubbles. (For more examples of acids and bases check out the pH scale in the back of the book!) It kept creating bubbles, which added air inside the

MAKE A HYPOTHESIS

→ Double the ingredients and try that again (you can use the same paint can). Do you think your paint can will shoot up quicker or slower this time?

7. Have your adult quickly flip over the paint can and step back.

Depending on your paint can, this should take anywhere from 1 to 20 seconds. At that time, your paint can will shoot straight up 10 to 20 feet (3 to 6 m) in the air.

Safety note:

Keep your eyes on that paint can because it's going to come down somewhere and you don't want it falling on your noggin!

TRY THIS ↓

Don't want your witch's brew to explode? Just use your paint can as a witch's cauldron! Pour in your baking soda just like before. Now, before you add your vinegar add a few drops of green food coloring to it and mix it up. Then add your green vinegar to your baking soda and watch as your witch's brew boils and bubbles over (with carbon dioxide bubbles, that is)!

paint can. That means the pressure was building and building and building . . .

And then all of a sudden, it exploded and shot right up!

That's right! The pressure was so great that the lid couldn't hold on anymore. The bottom shot upward because all that air that

was building inside the paint can rushed downward. Because of **Newton's Third Law of Motion** (for every action there's an equal and opposite reaction), that air moving downward pushed the bottom of the paint can upward.

→
(Make your own guess, then flip upside down to read!)

The more ingredients you use, the more bubbles you'll make, which will make everything happen faster. The pressure will build faster, and your paint can will shoot up even faster!

Rubber Bone

What are our bones made out of?

Materials
↓

Chicken bone
(the thinner bones work the best; a thicker bone may need more time in the vinegar)

A cup

Vinegar

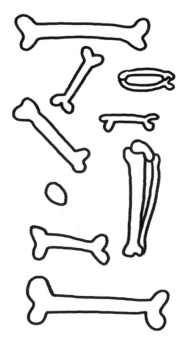

ABOUT THE EXPERIMENT

The bones in your body are made up of a framework of a soft and hard material. The softer material is a protein called collagen that allows our bones to be flexible. The harder material is made up mostly of calcium phosphate, which helps our bones stay strong and move properly. Sometimes, as we get older, our bones can lose calcium and get weaker. That's why it's important to eat foods that are rich in calcium and vitamin D (which helps our bones absorb the calcium).

In this experiment, we're going to fast-forward the aging process on a bone and remove the hard calcium altogether. Sounds spooky, doesn't it?

INSTRUCTIONS

1. Clean off your chicken bone with water so that all that is left is bone.
2. Place the bone in a cup and fill it with vinegar.
3. Wait 3 to 5 days.

As each day passes, you should feel your chicken bone getting more and more rubbery. By the end of the fifth day, you should be able to bend it in half without it breaking (this works best with thinner bones).

THE SCIENCE

 Whoa, what happened to the bone? It's all rubbery now!

 The vinegar dissolved the calcium throughout the bone. Calcium is what makes bones strong. And what's left is mostly collagen, which is soft, rubbery, and flexible.

 All you skeletons better stay away from vinegar then!

 Oh, definitely! If bones didn't have calcium, they wouldn't be strong enough to hold us up. They'd be very wiggly and rubbery, just like this chicken bone!

WHO TO KNOW
↓

Alice Ball was an African American chemist who created the first successful treatment for a very scary disease called leprosy, an infection that affects the skin and bones. She used the oil from a tree and found a way to make it injectable and absorbable by the body, which helped many people who were sick from the disease.

GLO
GLO
GLO

Glow-in-the-Dark Paint

Why does a black light look...purple?

ABOUT THE EXPERIMENT

I f you've ever seen a black light, you know that they have this purply looking glow to them. If it's a black light, then why isn't it black? Well, a black light emits a very special type of light called **ultraviolet light**. We can't actually see this type of light with our human eyes, but its wavelength is slightly shorter than violet light. Because ultraviolet light is so close to violet light, black lights often emit a little violet light too—giving black lights their purple glow! (To learn more about different types of light, check out the Electromagnetic Spectrum in the back of the book!)

This experiment will make use of that special type of light. If you're looking for a fun, messy group activity, then this is a great one! If you're able to fill a room with black lights and have everyone wear white, this would make for a super fun glow party—like the one we had on *Emily's Wonder Lab!*

Materials
↓

2 cups (480 ml) water

Bowl

Pliers

Yellow highlighter
(grab other colors, too,
if you like)

Spoon

¼ cup (35 g) cornstarch

Paintbrush

White paper
(or a white T-shirt)

Black light
(you can buy small black
light flashlights online)

INSTRUCTIONS

1. Add the water to your bowl.

2. Have your lab assistant (an adult) use the pliers to take off the bottom of the highlighter.

3. Shake the inside of the highlighter out.

4. Squeeze all of the highlighter liquid into the bowl of water. Stir with a spoon.

5. Add the cornstarch to your bowl to thicken up your glow-in-the-dark paint. If you want your paint to be thicker, add a couple more spoonfuls and stir with your spoon.

6. Now use your paintbrush to paint with your glow-in-the-dark paint. (I recommend painting in the dark with the black flashlight on! This makes it more fun and it's easier to see what you're doing.)

You may not see much with your regular lights on, but after you turn those off and turn your black light on, the paint should glow in front of your eyes. If the glow is dim, you may need more highlighter in your paint, or a darker room, or a stronger black light.

THE SCIENCE

 I didn't know highlighters could glow like that! How?

 Highlighters contain **fluorescent** chemicals. These **chemicals** have a special reaction to ultraviolet light—that's what your black light is emitting. When the ultraviolet light hits your fluorescent paint, it produces more visible light. Regular paint can't do that.

 Black lights are cool! What makes that ultraviolet special?

 Ultraviolet light has a shorter wavelength than blue or purple light and it also has more energy.

 Whoa, so it adds energy to my paint?

 Exactly! It energizes the fluorescent chemicals in your paint, which makes it glow with visible light we can see with our eyes!

→

Marie Curie was a scientist who also worked with things that glow. But her work glowed for a different, more dangerous reason: It was radioactive! Marie Curie helped us learn about radioactivity and was the first woman to win the Nobel Prize for her groundbreaking research.

NOTES!

Glowing Lava Lamp

How do huge cruise ships float on top of water?

ABOUT THE EXPERIMENT

Ever wonder how large boats—like a 200,000-ton cruise ship—can float on top of water? It's all about **density**! The boat sure does weigh a lot, but that boat is less dense than water (meaning, if you had a cruise ship–sized container of water, the water would weigh more than the cruise ship!). Sure, the boat has a lot of metal and other heavy things inside, but it also has a lot of empty space. This is why the cruise ship floats on water—things that are less dense than water will float on top of it. In this experiment, we will use the principle of density to create a beautiful lava lamp. This is a fun one to do in the evening when you can make the room really dark and watch the glow work its magic. Or just shut yourself in the bathroom and turn off all the lights! →

Pliers

Yellow highlighter

Glass

Water

(enough to fill three-quarters
of the glass)

Cooking Oil

(enough to fill the
other one-quarter of the
glass; I use olive oil)

Black light

(you can get small black
light flashlights online)

1 small bowl of salt

(you can have more or less,
depending on how long you
want to do this experiment)

INSTRUCTIONS

1. Have your lab assistant (an adult) use the pliers to pull off the bottom of the highlighter.

2. Shake the inside of the highlighter out.

3. Fill your glass three-quarters of the way up with water.

4. Squeeze the inside of your highlighter into the water, emptying all the contents into the water.

5. Fill the rest of your glass with oil.

6. Make sure you're in a dark room. Then turn on your black light and place it under your glass.

7. Little by little, toss pinches of salt on top of the oil.

As you toss in your salt, you should see some globules of oil fall down to the bottom and then just seconds later rise back up to the top.

THE SCIENCE

Wow! The oil is moving down and up and down and up, each time I add salt.

That's right! Salt is denser than oil and water and so it sinks to the bottom of your glass and brings a little bit of the oil down with it. Once it gets to the bottom of the glass it has time to dissolve into the water, letting go of the oil and allowing the oil to rise back up to the top.

Because the oil is less dense than water?

That's right!

NOTES!

_____

_____

_____

_____

_____

_____

_____

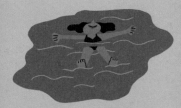

You can float on top of water in the Dead Sea. The Dead Sea is a very salty body of water bordering the country of Jordan. People travel to the Dead Sea to float right on top of it! Well, you can float higher than you can in regular water, anyway. That's because the water is so salty and so much more dense than freshwater or even normal ocean water.

Why does it glow?

Highlighters contain **fluorescent** chemicals. The black light emits **ultraviolet light** that interacts with those chemicals and gets them really excited, which ultimately causes them to glow in the dark. (Be sure to check out the Electromagnetic Spectrum in the back of the book!)

Invisible Ink

Why do some things glow under black light?

ABOUT THE EXPERIMENT

If you've ever been in a room with a black light, you've noticed that some things (like parts of your clothing or even your teeth) glow while other things don't. This is because things that contain something called **phosphors** will glow when exposed to the special type of light that black lights emit called **ultraviolet light** . (Check out the Electromagnetic Spectrum in the back of the book!) Phosphors can be created naturally (like the kind in your teeth) and they can also be human made. When companies want paint or stickers to glow under black light, they add **chemicals** that contain phosphors.

In this experiment, we'll make a secret message by breaking down the natural phosphors in paper. This one is super fun to leave secret messages to friends and family. Because you'll be working with bleach, you'll need your lab assistant (an adult).

DID YOU KNOW?
↓

Scorpions naturally glow under black light. This is because their **exoskeleton** (the skeleton on the outside of their body) contains phosphors. So, if you are worried you might be in an area with scorpions, be sure to carry a black light with you. You'll be able to spot them much more easily.

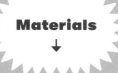
INSTRUCTIONS

1. Dip the tip of your paint-brush into the bleach—you don't need much!

2. Turn on your black light and point it toward the paper (it may be helpful to have a friend do this while you write the message).

3. Write your top-secret message with your paintbrush, dipping your paintbrush in the bleach when you need more.

4. Let your message dry (5 minutes or so).

5. Entrust a friend with your black light to reveal your hidden message.

Your friend should see the message you wrote in bleach turn up as grayish black while the paper glows under the black light.

THE SCIENCE

 The black light revealed my message!

 That's right! But is your message glowing?

 Actually, no. The paper seems to glow, but my message looks kinda . . . gray.

 Exactly! Your computer paper is fluorescing ever so slightly. That means that it contains phosphors that take energy from ultraviolet light, and it uses that energy to emit visible light back to your eyes. But the bleach breaks down some of those fancy fluorescent chemicals. This means that the parts where you added bleach won't fluoresce under black light anymore.

 That's so cool! And when you turn the black light off, you can't see the message at all.

 That's right! Because bleach is clear to our eyes in regular daylight, it makes it nearly impossible to see where you wrote your message!

EGGS Science

Scie

ellent
nce

Egg in a Bottle

Lab Assistant (an Adult!) Required

What makes my ears pop?

ABOUT THE EXPERIMENT

When we reach higher altitudes, our ears "pop"—it's all about air **pressure**! Think of the air as little fists that are punching everything around you all the time. When you have a high-pressure environment, those fists punch harder and more frequently. When you have a low-pressure environment, it's like they're punching in slo-mo. They're less strong.

We have little air pockets inside our ears. When the air pressure inside your ears matches the air pressure outside, you don't feel anything. But if you move up a hill, or fly up in a plane, or take an elevator to a really high floor, the atmosphere gets thinner and the pressure gets lower. When that happens, the air pressure inside your ears is much **STRONGER** than the air pressure outside of it.

This increased air pressure stretches your eardrum—you can actually hear the difference (the world may sound a little muffled). But eventually, the little fists inside your ear are able to escape and equalize the pressure inside your ear to match the outside. That's when you hear a POP! So every time your ears pop, you know that air is either rushing out of your ear (when you go to a higher altitude) or rushing into your ear (when you go back down to a lower altitude).

This experiment will use the power of air pressure to do something really neat—get an entire egg inside of a bottle!

Materials

↓

Scissors

Coupon paper

(the kind you get in your mailbox)

Lighter

Glass bottle

with an opening slightly smaller than the width of an egg (I use glass milk bottles)

2 eggs, boiled and peeled

Safety note:
To do this experiment, you'll need to light something on fire, which requires adult supervision.

INSTRUCTIONS

1. Using your scissors, cut out a 4-by-2-inch (10-by-5-cm) rectangle from your coupon paper.

2. Fold that paper in on itself to create a little tube that fits inside your bottle.

3. Have an adult light the top of the paper on fire and throw it in your bottle.

4. Very quickly place your egg on top of the bottle.

You should see your entire egg very quickly get pushed into your bottle! If it didn't work, the fire may have gone out too quickly for the reaction to happen. Try it again and this time make sure the coupon paper is still on fire when you put your egg on top of the bottle.

THE SCIENCE

 That happened so quickly! The entire egg is inside the bottle. How'd it get pushed in?

 When you put your fiery coupon paper inside the bottle, it heats up all the air inside. When air gets hot, it wants to take up more space, so many of those air molecules danced right out of your bottle. Then, when you put your egg on top, it prevented the flame from getting oxygen, so it went out. Now you have less air in the bottle than you did before and that air is cooling down. When air cools down, it wants to take up LESS

space and condense together. That means the air pressure inside your bottle is now a lot less than the air pressure outside. The air pressure outside forces your egg in super quickly—and just like that, you have an egg in a bottle!

 Wow! Okay, so now how do I get it out . . . ?

 That's the million-dollar question! There are a few ways to do it, but I like to use a knife to cut it up, and then I shake out the egg guts.

49

Allergic to eggs or don't have any on hand? Don't worry, you can do this same experiment with a small balloon filled with a little bit of water! Add a little water to a balloon and then blow it up so that it is slightly larger than an egg (and larger than the opening of your bottle). Tie off the top. Now, repeat the experiment on page 49 and just use your balloon as your egg!

NOTES!

Walking on Eggs

How do bridges hold so much weight?

ABOUT THE EXPERIMENT

Did you know that arches are the most common shape found in bridges? This is because an arch is one of the strongest shapes in engineering. It helps distribute a lot of weight across a larger area, so that no one point in the bridge has to hold all the force. This makes the bridge stronger and able to hold more weight.

With this experiment, you'll get some hands-on (or maybe FEET-on) knowledge of the power of the arch. This can get a little messy, so it's a great one to do outside. To be especially safe, do this on the grass, or have friends hold your hands on each side. On *Emily's Wonder Lab*, we filled an entire room with a few thousand eggs, and I challenged my little scientists to walk across the room without breaking any. What do you think . . . could you do that? Are you up for the EGG CHALLENGE? →

1 carton of 12 raw eggs

(or 2 cartons if you want to use both feet!)

INSTRUCTIONS

1. Place your carton of eggs outside, in the grass if you can.

2. Open up the carton so that the eggs are uncovered.

3. Hold on to something nearby (maybe a friend's hand) and slowly use one foot to step on top of the carton of eggs while keeping your other foot in the air.

4. If you have two cartons, now use the second foot to step on the other carton of eggs.

5. Let go of your friend's hand (or whatever you were using to balance).

You should be able to stand on top of your carton of eggs without breaking any (or, without breaking many!).

THE SCIENCE

 Did you break any eggs? If you did, it probably wasn't many, right?

 Only one cracked! But I thought eggs were fragile?

Eggshells are made of a fragile material, but it's the shape of the egg that gives it its strength. An egg is basically two arches that come together in the middle. When force is applied to the egg, the arch spreads out that force down and along the arch. This shape allows a hen (or an ostrich, or any other animal that lays eggs) to sit on their eggs without breaking them!

MAKE A HYPOTHESIS

→ **What would happen if you only stood on one egg?**

→ **Test it out!**

→ (Make your own guess, then flip upside down to read!)

Norma Merrick Sklarek was the first Black woman to be licensed as an architect in California and New York. She worked on the design of the U.S. embassy in Tokyo and Terminal One at the Los Angeles International Airport, among other incredible projects. She succeeded while overcoming barriers of discrimination and despite not having any mentors, and said, "In architecture, I had absolutely no role model. I'm happy today to be a role model for others that follow."

NOTES!

_____

_____

_____

_____

_____

_____

_____

_____

_____

Are your feet gooey with egg guts? The reason why that may not have worked is because when you're standing on many eggs at once, you're distributing your weight across each of those eight eggs. So, if your foot can touch eight eggs, then your weight is divided among those eight eggs. That makes it much easier to hold you up without breaking. An egg is strong, but it's not invincible! If you put all your weight on one egg, it might just break.

Naked Egg

What's an eggshell made of?

ABOUT THE EXPERIMENT

We're going to reveal what's under the armor (eggshell) of an egg and create a naked egg! Did you know that eggshells are made up of the same stuff that seashells are made of? Seashells and eggshells are mostly calcium carbonate. In this experiment, we'll use vinegar to kick-start a **chemical reaction** with the calcium carbonate in the egg. It takes 24 to 48 hours for the reaction to occur, but believe me—it's worth the wait!

INSTRUCTIONS

1. Fill your glass with vinegar (you just need it to be high enough so that it will cover an egg).
2. Carefully place your raw egg in the glass.
3. You want your egg to be completely submerged in the vinegar—help it stay under by placing your spoon on top of the egg.
4. Let's observe what's happening at this point. Do you see bubbles? What do you think is causing that?
5. Now, wait 24 to 48 hours . . .
6. Let's observe what's happening at this point. What do you see in the glass? Is it . . . murky? Brown? Gunky? What do you think that is?
7. Very gently wash your egg under running water in the sink. Use your thumb to gently brush off the rest of the shell.

You should be able to brush off the rest of the eggshell with this method. After this, your egg will feel rubbery and look slightly more see-through (you won't actually be able to see through the egg, but it will look slightly more transparent than it did with the shell on).

Materials
↓

Large glass

1 to 2 cups (240 to 480 ml) vinegar

1 raw egg in the shell
(any kind will work)

Metal spoon

THE SCIENCE

 What were those bubbles when I put the egg in the vinegar?

 That was the chemical reaction happening before your eyes! Vinegar reacts with calcium carbonate to dissolve the shell and create carbon dioxide bubbles.

 After a day, the vinegar got so murky looking and there was a bunch of brown stuff in the glass. Gross! What was that?

 That's part of the shell that was removed from the egg.

 If the shell is removed . . . how is the egg still together?

 Well, under the eggshell, an egg has membranes—an outer membrane and an inner membrane. These hold all the guts inside the egg.

 How strong are they? Can I bounce one?

 It's worth a try! →

55

Hold your naked egg 1 inch (~2 cm) above a hard, flat surface—and let go! Did it bounce? Try 2 inches (~5 cm) . . . then 3 inches (~7 cm)! See how high you can go before it . . . SPLATS! Warning: This can get messy! The height you are able to bounce your rubber egg will depend on the strength of the membrane of your particular egg. How high did you get? I once bounced a naked egg 12 inches (~30 cm) off the ground before it broke!

NOTES!

Rainbow Bouncy Egg

Why do my fingers get wrinkly after taking a bath?

ABOUT THE EXPERIMENT

Have you ever looked at your fingers after taking a bath? They look wrinkly, don't they? Did you know the reason they are wrinkly is because water molecules from the bath have actually entered your skin's cells? They marched right into the upper layer of your skin and stretched out the area a little bit, making your fingers and toes look really bumpy and wrinkly! This happens because of something called **osmosis**. In this experiment, we're going to take the naked egg experiment from page 54 up a notch by making them RAINBOW through the power of osmosis. →

Materials

5 cups (1.2 L) vinegar

5 drinking glasses

Food coloring in 5 colors

(I used red, orange, yellow, green, and blue; you'll need about 10 drops of each color.)

5 raw eggs in the shell

Spoon

INSTRUCTIONS

1. Place 1 cup (240 ml)of vinegar in each of your five glasses.

2. Place 10 drops of red food coloring in one glass, 10 drops of orange in another, 10 drops of yellow in the next, 10 drops of green in the fourth glass, and then 10 drops of blue in the last one.

3. Use your spoon to mix the colors in the vinegar.

4. Gently place a raw egg in each of the five glasses.

5. Wait 24 to 48 hours.

6. Gently run your egg under water and wash off the rest of the shell.

Each of your eggs will be a different color and it should look like the color is inside the egg itself.

THE SCIENCE

 Look at my beautiful rainbow eggs! Wait, are they colorful on the inside too?

 Yes!

 How did the color get in there?

 Through osmosis! Osmosis is all about the movement of water through a semipermeable **membrane**. Water molecules are so small that they can get through things that other molecules can't—like the membrane of an egg.

Water marched right on over from the vinegar through the membrane and into the egg.

 But why does water want to do that? What made it move?

 With osmosis, water likes to move from an area of high concentration of water to an area of low concentration of water. It's kind of like if you had two rooms separated by a swinging door. In one room you had a huge party of water molecules, and in the other you just had a couple water molecules having a small, boring party. Water molecules like things to be fair, so some of them will move through the door from the

58

The reason gargling with salt water makes your sore throat feel better is because of the power of osmosis. When you have a sore throat, the cells in your throat are swollen and filled with liquid. When those cells are exposed to salt water, some of the water molecules inside will march right out of your cells into the salt water. This helps to temporarily reduce swelling in your throat. It may not fix the problem, but it sure makes it feel better for a little while. Thanks, osmosis!

big party to the small party to even things out a bit!

And because we dyed the water different colors, the food coloring went inside the egg too?

Exactly! And that's how you get colorful eggs!

NOTES!

_____

_____

_____

_____

_____

Growing and Shrinking Egg

How do plants drink water?

ABOUT THE EXPERIMENT

Have you ever wondered how plants drink water? Well, the main way roots soak up water molecules is through the process of **osmosis**! When a plant is "thirsty" and the soil around the plant's roots is filled with water, the water molecules can march right on through the plant's cell membranes and into its roots, which then help it get to the rest of the plant. With this experiment, we're going to put the process of osmosis to work. We'll take our naked egg that we created in the experiment on page 54 and make it shrink and grow using the science of osmosis!

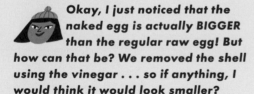

Materials

1 naked egg

(see page 54)

1 raw egg

(used for comparison)

1 cup (240 ml) honey or corn syrup

Glass or cup

1 cup (240 ml) water

INSTRUCTIONS

1. Take your naked egg and compare it to a normal raw egg—what do you notice?

2. Pour the honey into the glass. Now place your naked egg in the honey for 24 to 48 hours.

3. You should see your egg shrink slightly. The membrane around the egg will appear a little wrinkly and the inside of the egg will look slightly darker, like the color of honey (or corn syrup).

4. Now place the egg in a cup of water for 24 to 48 hours.

After the last step, your egg should have grown slightly and the membrane should be more stretched out and look less wrinkly.

THE SCIENCE

Okay, I just noticed that the naked egg is actually BIGGER than the regular raw egg! But how can that be? We removed the shell using the vinegar . . . so if anything, I would think it would look smaller?

Great observation! It got bigger because of osmosis! More water molecules entered through the semipermeable **membrane** and filled the egg to make it just a little bit larger. The water molecules marched inside because the water concentration outside the egg was higher than it was inside the egg. The egg swells because there is now more water inside. What happened to the egg after you soaked it in honey? →

VINEGAR, THEN CORN SYRUP

RAW EGG

VINEGAR

It shrank! Ewww, the membrane looks all shriveled, and the inside is a lot yellower now. How'd that happen?

 Well, take a look at your cup of honey . . . notice anything there?

 It's all watery now! Where did that water come from?

 From inside the egg! Osmosis worked the other way around this time. The water concentration inside the egg was much higher than it was outside the egg. So, this time, the water molecules marched right on out into the honey, making the honey in the cup watery and removing a lot of the water that was inside your egg.

 And I can just keep doing this, right? Because when I put it back in water overnight, it got bigger again.

 Exactly! Well, as long as those egg membranes don't break. Osmosis is a very cool thing!

TRY THIS
↓

Let's do the experiment again, but this time with three naked eggs (see page 54) and three different cups of water. The first cup has no sugar in it. The next cup has ¼ cup (50 g) of sugar. The last cup has 2 cups (400 g) of sugar dissolved in the water. Let the eggs sit overnight. You should see osmosis at work here! After 24 hours, the egg in the cup with plain water shouldn't look too different, the egg in the second cup should look a little smaller, and the egg in the last cup should be the smallest of all. This is because the cups with sugar had lower concentrations of water in them (so the water wanted to march from the egg into the cup). The cup with the most sugar had the lowest concentration of water, which is why more water left the egg in that last one than in the first two cups.

NOTES!

_____

_____

_____

_____

BiRTHDAY SO CREAT

VINEGAR

DAY CIENCE IONS!

DIY Ice Cream

How is ice cream made?

ABOUT THE EXPERIMENT

This is one of my favorite experiments because it always turns out so much more delicious than you'd expect! Let's learn how you can make ice cream right at home. It's actually pretty similar to how professionals make it. Have you ever been to a rolled ice cream store where they make the ice cream in front of you? They pour their liquid ice cream mixture on a really cold plate to supercool the ingredients. That transforms it from a liquid to a solid. I'm going to show you how to supercool your ingredients at home. Go ahead and get creative. Add sprinkles! Chocolate syrup! Peanut butter chips! Food coloring! Pickles! Any flavor is possible!

INSTRUCTIONS

1. Fill the sandwich-size bag with the milk, half-and-half, sugar, and vanilla and seal the bag tightly.

2. Add the salt to the gallon-size bag filled with ice.

3. Add the sealed sandwich bag to the large bag.

4. Seal the large bag and try to make sure there's not much air in there.

5. Put on your winter gloves and shake and massage the bag (which is now surrounded by ice) for 5 full minutes. Woo-hoo! That's a workout, isn't it? Shake, shake, shake! Who wants ice cream?! WE DO! Shake, shake, shake!

6. Carefully open up the larger bag to retrieve the sealed sandwich bag.

7. Use a spoon to scoop out your ice cream from the bag into a bowl and enjoy your ice cream—made with science!

Your ice-cream mixture should have transformed from a liquid to more of a solid. If it didn't work, perhaps you need more ice and salt in your outer bag. Make sure that the inner bag is sealed tightly so no salt creeps in!

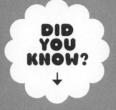
Salt was once used as money! Today we just think of salt as a great seasoning, but historically it was also used as a way to preserve food. This could be lifesaving. You could put salt all over meat and instead of it going bad in a day, it could last for weeks or even months! Salt helps soak up the moisture in the meat, which bacteria need to grow. Bacteria (like salmonella) will make you sick and can even be deadly. Salt allowed people to keep their food safe for longer, and was so valuable it was even used as currency.

THE SCIENCE

How did my milk become ice cream?

You supercooled your milk into ice cream using science! The salt in your larger bag lowered the freezing point of ice, which made a lot of that ice melt quickly. The melting ice needed a lot of energy from its environment to be able to melt so quickly. The energy that it took was HEAT energy. So, by the process of melting, it made the whole environment colder. We needed that bag to be really cold to transform the milk into ice cream, and SALT was the key ingredient that made it happen.

IT'S ALL THANKS TO ME! YOU'RE WELCOME!

LED Birthday Cards

Why do tall buildings have metal rods on top?

ABOUT THE EXPERIMENT

This experiment will show you the power that metal has when it comes to electricity! Next time you're in a big city, look at the tops of the buildings. You'll probably find a few tall metal rods. Those rods help protect the building from lightning. Instead of hitting the building structure, lightning will strike the rod, which goes all the way into the ground. This will direct all of the **current** from the lightning away from the building and down into the ground.

This works because different types of metal are good **conductors** of electricity. This means that electrons can move through them really easily. Copper, in particular, is a great conductor because there are free-flowing electrons moving throughout it. It's like a superhighway for electrons!

This experiment is so much fun because you can get really creative! It's perfect for holidays, birthdays, or ANY occasion. This one requires a little patience and can take 15 to 30 minutes to complete, depending on how much you want to decorate your card.

Materials
↓

Paper
(cardstock or computer paper works great)

Markers

LED light
(you can buy this online or at a hardware store)

Pencil

Copper tape
(you can buy this online or at a hardware store)

Scissors

Clear tape

Watch battery

INSTRUCTIONS

1. Fold your paper in half hamburger style.

2. Decorate your card.

3. Use your pencil or the leg of the LED to poke a tiny hole where you want your LED to be.

4. Now open your card.

5. On the left side we'll draw our circuit. Using your pencil, outline your circuit. The final circuit will match the drawing here. Your LED location (the small red circle) may be different, depending on your design.

6. Fold the bottom left corner of your card and make sure that your circuit will connect.

7. Get your copper tape. We want to cut off a piece that matches the length of the left side of your circuit (the "L" shape). You want it to be one continuous piece.

8. After cutting your copper piece, carefully remove the strip on the back and lay the copper sticky-side down over your outlined circuit on the left side. Simply bend it over on itself at the corner of your L. Make sure you use one continuous piece of copper tape here! If you use two pieces to create the "L" shape, it won't work because the sticky part will block the connection. The copper needs to touch the entire way from the LED to the battery.

9. Now measure a copper strip for the right side, remove the backing, and place it sticky-side down over your outline. Don't cover the hole for your LED!

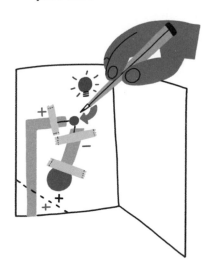

10. Place your LED in the hole with the light on the outside of the card. Place the longer leg of the LED on the positive end of your circuit and the shorter leg on the negative end.

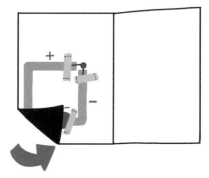

11. Use your clear tape to tape down the legs on top of the copper tape.

12. Place your battery with the positive side (you'll see a "+" symbol) facing up, touching the positive side of your circuit.

13. Use your clear tape to tape it down (don't cover the whole battery).

14. Fold over the corner to connect the circuit.

When you fold over the corner and press it down, the copper tape should come into contact with your battery and your LED should light up. If it doesn't, make sure your LED legs are in direct contact with the copper tape and your battery isn't fully covered by your clear tape. ➜

NOTES!

_____

_____

_____

TRY THIS
↓

Place the LED legs on either end of your watch battery. Now switch it, with the longer leg on the other side of the battery. Notice how it only works in one direction? This is why it's so important to lay out your circuit correctly, because in your LED, electrons can only flow one way. So, you need to match the positive end of the LED to the positive side of the battery and the negative end of the LED to the negative side of the battery. Once you do this, your light will glow!

THE SCIENCE

 It lights up!
What is happening?

 Your creation is beautiful! The LED only lights up when the electricity is flowing through it. When you press that corner down, you are making a circuit. In the circuit, electrons move from one side of the battery, through the LED, and back to the other side of the battery. The electricity flows around and around the circuit only when there is a completed path from one side of the battery to the other.

 You mean I created my own circuit with just
tape, a battery, and this little light?

 Yes, you did! The battery is providing the movement of electrons, the copper tape is giving them a highway to run through, and your LED is using their movement to create light. And when you don't push the corner down, you have an open circuit, which means electrons are not moving through the LED and the light doesn't turn on.

 So, I can turn it on and off, on and off, on and
off all day long, just by pushing this button?

 Exactly! Well . . . until your battery runs out of juice. But that won't happen for a long time!

Self-Inflating Balloon

If you accidentally let go of a balloon outside, what happens to it?

ABOUT THE EXPERIMENT

If you let go of a helium balloon outside, it will rise and rise and eventually you won't be able to see it anymore. As the balloon gets higher in the atmosphere, the air **pressure** outside of the balloon gets lower, which means the air molecules are more spread apart. The air inside the balloon wants to match the air pressure outside the balloon, so the air molecules inside the balloon start to spread apart too. This makes your balloon stretch bigger and bigger. Eventually, the balloon can't get any bigger and it will POP, send-

ing the balloon parts back down to the ground. Sometimes these balloons can end up in the water or in animals' habitats, so it's very important to hold on to your balloons tightly when you're outside!

This experiment will stretch out a balloon and make it bigger too. But instead of launching it in the air, we're going to add air inside the balloon using a chemical reaction. →

Balloon

2 tablespoons baking soda

Plastic bottle

¼ cup (60 ml) vinegar

DID YOU KNOW?
↓

Students can send science experiments to the edge of space using balloons. These are a special type of balloon called high altitude balloons. Many are larger than a car when they are launched from the ground. These balloons can give us information about the weather and other scientific information about our upper atmosphere!

INSTRUCTIONS

1. Blow up your balloon once and then let the air out (this will stretch out your balloon).

2. Pour the baking soda into the bottle.

3. Hold your balloon open while your friend or lab assistant (an adult) pours the vinegar inside your balloon. Try to pour in as much as you can!

4. Very carefully, trying not to spill the vinegar, place your balloon on the opening of your bottle. Make sure there is no gap between the balloon and the bottle.

NOTES!

_____

_____

_____

_____

5. Lift your balloon upward so that the vinegar mixes in with the baking soda.

6. Shake your bottle around so that the baking soda and vinegar fully mix.

As soon as you lift up your balloon, you should see bubbles form as the vinegar and baking soda begin to mix. Those bubbles will almost instantly make your balloon expand and stand straight up! Your balloon may not completely fill up with air—it all depends on how much is in your bottle.

TRY THIS
↓

Try adding fewer ingredients to your bottle and balloon. How does that change how much your balloon inflates? Less vinegar and baking soda will lead to fewer carbon dioxide bubbles, meaning your balloon won't blow up as big.

THE SCIENCE

 Whoa! The balloon filled with air out of NOWHERE! How'd that happen?

 Let me show you. I want you to try something. Close your mouth. Now blow air into your mouth and cheeks. What happened to your cheeks?

 My cheeks puff out.

 Exactly! You added more air to your mouth from your lungs. That increased the air pressure inside, and that pressure pushed out on your cheeks. Something very similar is happening in your bottle. Instead of your lungs pushing out more air, the chemical reaction is doing it! The baking soda and vinegar create carbon dioxide gas. That gas is adding to the air molecules that were already in the bottle. Those air molecules don't have anywhere else to go, so they leave the bottle and fill the balloon. As more gas is created, the air pressure increases and that pushes out on the balloon, filling your balloon in the process!

DIY Fire Extinguisher

Lab Assistant (an Adult!) Required

What's in a fire extinguisher?

ABOUT THE EXPERIMENT

There are a few different types of fire extinguishers, but many of them work by throwing carbon dioxide (CO_2) on a fire. The fire extinguisher holds carbon dioxide gas under **pressure** and when you press the handle, you release a lot of carbon dioxide at once. CO_2 under intense pressure can freeze, so the air that comes out will be really cold. Air contains water as a gas (water vapor). In this form it's transparent. When it cools down, the water vapor turns into tiny droplets of water. That fog you see is the water vapor condensing into teeny tiny water droplets because the CO_2 around it was so cold. Here's the really important part: CO_2 is **denser** than air, so it will sink and sit on top of the fire, snuffing out the flame!

In this experiment, we're going to make our own fire extinguisher. You can even use it to put out your birthday candles!

Materials
↓

Lighter

Candle

¼ cup (57 g) baking soda

¼ cup (60 ml) vinegar

Glass

Safety note:
To do this experiment, you'll need to light something on fire, which requires adult supervision.

INSTRUCTIONS

1. Recruit your lab assistant (an adult!) to light your candle.
2. Mix the baking soda and vinegar in your glass.
3. Wait until the bubbles subside.
4. Bring your glass close to the candle and start to tilt your glass over the flame, as if dumping the air inside on top of your candle (just don't let the liquid fall out).

After you tip your glass over the flame, you should see the flame move a bit as if there is wind in the room and then go out!

TRY THIS
↓

Try the experiment again, but this time after step 3, pour your air into another glass and then use that glass to pour over your candle. Did it still go out? This helps prove that even though you can't see the carbon dioxide in the glass, it's still there!

THE SCIENCE

 Whoa! The candle went out! Why? I didn't see anything come out of the glass!

 You may not have seen anything, but you definitely poured something on top of that flame! When you mixed the baking soda and vinegar, you created carbon dioxide bubbles in an acid-base **chemical reaction**.

 So, the carbon dioxide put out the flame? But how?

 Well, you know how a flame needs oxygen to burn? Carbon dioxide is denser than air, so it sinks. That means it'll fall out of your glass kind of like water would. It sat on top of your flame like a blanket and starved it of the oxygen in the air. Once your flame couldn't get enough oxygen, it just went out!

DID YOU KNOW?
↓

For really big wildfires, firefighters can use satellite data to help track where the fire is spreading. Satellites can see our planet in infrared light, a special type of light emitted by hot objects. (To learn more, check out the Electromagnetic Spectrum in the back of the book!)
By looking at infrared light, this allows firefighters to see through smoke like Superman and find where the fire is spreading on the ground.

Balloon-Popping Oranges

What gives oranges that delicious smell?

ABOUT THE EXPERIMENT

Have you ever smelled the peel of an orange? Notice how it has that wonderful citrus smell? That's a specific type of **chemical** that we're going to use to pop our balloon in this experiment! Although I will say that I HATE the sound of popping balloons, so this experiment always scares me a little—but it is fun to do!

MMM, SMELLS NICE!!!

Materials
↓

A cheap balloon

(This is important! Get a balloon that looks really clear when you blow it up. Water balloons work. Some of the nicer, thicker balloons you find online won't work because they are chemically made of a different material.)

An orange

INSTRUCTIONS

1. Blow up your balloon and tie it off.
2. Cut off the peel of the orange.
3. Squeeze the peel over the balloon so you get a drop of the juice on the balloon.

Within a second of squirting the orange peel onto the balloon, the balloon should POP. If it doesn't, try squirting more of the orange peel juice onto the balloon. If it still doesn't pop, you may have the wrong type of balloon.

THE SCIENCE

 Ah! The balloon popped almost instantly! What's in that orange?!

 I know, right? It always scares me! Orange peels have a chemical called limonene in them. This is the chemical that gives the orange peel that wonderful citrus smell!

 Mmm, it smells delicious! But why is it so deadly to the balloon? Should I be eating this stuff?

 It's all about chemistry! Some chemicals are perfectly safe in the right quantities for us, but not so safe even in really small quantities for other materials. Limonene causes the rubber in the balloon to dissolve a little. This creates a weak point in the balloon and . . .

 POP!

 It scares me every time!

TRY THIS
↓

Try popping the balloon with a lemon or lime! Those fruits also contain the chemical limonene in them, so they will work to dissolve the rubber and pop the balloon as well. Limonene is just more concentrated in oranges, so oranges usually work the best.

Save the Balloons

Why does it take water so long to boil?

ABOUT THE EXPERIMENT

This experiment will show us exactly why it takes water so long to boil. Water has something known as a "high heat capacity." This means that it takes a lot of energy to heat it up. Basically, water is stubborn and needs a lot of motivation to change its temperature! This means you have to be really patient while the heat on your stove slowly works to heat up all of the molecules in your pot of water.

In this experiment we'll show off the **heat capacity** of water. This is a fun experiment to do with some leftover birthday balloons! You can show your friends how to save those beautiful balloons from popping with just a little bit of water . . . and science!

Materials

↓

2 latex balloons

Lighter

Candle

Water

Safety note:
To do this experiment, you'll need to light something on fire, which requires adult supervision.

INSTRUCTIONS

First, let's watch what happens when we put a balloon over a flame!

1. **Blow up one balloon and tie it off at the end.**
2. **Have your lab assistant (an adult!) light the candle.**
3. **Place the balloon about 1 inch (~2 cm) above the candle flame.**

You should have seen your balloon POP!

Now, let's try that again, but this time with water inside your balloon.

1. **Place your second balloon under a running faucet and fill it up with about ½ cup (120 ml) of water (don't worry about being too precise).**
2. **Carefully blow up your balloon like you did before and tie it off at the end.**
3. **Place it about 1 inch (~2 cm) above the candle flame like you did before.**

THE SCIENCE

 My balloon without water popped almost immediately!
What happened?

 The heat from the flame started exciting the molecules in your latex balloon. This makes the molecules in the latex wiggle a bit and that makes the bonds between them weaker. At a certain point they get so wiggly from all the heat energy that the bonds between them break and BOOM—a tear is created and there goes your balloon! What happened to your balloon with some water in it? →

That one didn't pop! Why is this one different?

Isn't that cool?! It's because the water is literally saving your balloon! Remember how before all of that heat energy from the flame made the latex molecules wiggle, shake, and break their bonds? Well now that energy is being transferred to the water! It's saving your latex molecules like a friendly security blanket.

I CAN TAKE THE HEAT FOR YOU PAL!

  But why isn't the water boiling?

It's the same reason it takes water FOREVER to boil on the stove! It's stubborn and requires a ton of heat energy to boil. It may not seem like much, but that water can take a lot of energy without changing its temperature too much.

 Next, go ahead and just set your balloon with water in it down on top of your candle.

 Oh my gosh! The balloon didn't pop but the candle went out!

 Exactly! By placing your balloon on top of the candle you prevented the flame from getting the oxygen it needs and you snuffed it out! Try to lift your balloon off of the candle.

 It's stuck!

 That's right! When you placed your balloon over the candle, the little bit of air at the top was really hot. But after the flame went out, that air cooled down. When air cools down, it wants to take up less space and squeeze together more tightly. You've created a partial **vacuum** in there! The air pressure outside is so much greater than the air pressure inside that it's forcing your balloon inside the candle. You're going to need to use a little muscle to get it out! Once you do, look at the bottom of the balloon. Notice anything?

 It looks black! What is that stuff? Is that because the balloon burned?

 Actually, no. That's coming from your candle! Your candle is likely made up of paraffin wax, which contains hydrogen and carbon atoms. When a candle burns, it goes through **combustion**, creating light and carbon dioxide. But if it can't find enough oxygen, it won't transform the carbon into carbon dioxide, and you'll get some leftover carbon. You can sometimes see this as black smoke in your flame. That's the same stuff that's on the bottom of your balloon. We call this soot.

 Oh wow! I can rub it off with my thumb!

MY BUTT'S STUCK!

NOTES!

Magically Moving Bubble

Why do balloons make your hair stand on end?

ABOUT THE EXPERIMENT

When you rub a balloon on your hair, you're transferring electrons from your hair to the balloon. This does a couple of things. One, it makes the balloon have a **static charge**. Two, it makes your hair now have a charge that is *opposite* of the charge on the balloon, so it will become attracted to the balloon! Each hair also has the same charge, so they will repel each other— making them stand up even more!

In this experiment, we'll use static charge to actually move something. It looks like a magic trick if you don't know the science! This works best if you have a large, smooth surface like a countertop. Because you might end up getting soap on the floor, this is a great one to do in your kitchen.

Materials
↓

Dish soap

Glass of water

Spoon

Straw

Balloon

Head of hair

TRY THIS
↓

Let's try moving other forms of water. Turn on the faucet so you have a really thin stream. Rub the balloon on your head and place it next to the stream. What happened? What happens if you turn the faucet higher? Do you notice it move as much? (The heavier the water gets, the harder it is to move with the static.)

INSTRUCTIONS

1. Add a few squeezes of dish soap to your glass of water. Stir with a spoon.
2. Throw some water on a flat, smooth surface (about ½ cup [120 ml] should do).
3. Place the end of your straw on the surface and blow a bubble in the soapy water.

4. Blow up your balloon and tie it off.
5. Rub the balloon on your head for about 20 seconds.
6. Place the balloon in front of the bubble (about ½ inch [~1 cm]).
7. Repeat steps 5 and 6 as necessary.

You should be able to quickly move your bubble ever so slightly with your balloon. Make sure you get your balloon very close to the bubble so that it can feel the static charge.

THE SCIENCE

The bubble moved! I feel like I have a superpower!

That's right! By rubbing the balloon on your head, you moved electrons from your hair to the balloon. Your balloon temporarily will have a static charge.

But how does that static charge move a bubble?

Because in a bubble, atoms are free to move around, which makes them easily affected (and moved!) by static charge.

Self-Filling Balloon

Is there air on the moon?

ABOUT THE EXPERIMENT

HEY, NO FAIR, WHERE'S MY ATMOSPHERE?

Have you ever looked up at the moon and wondered if there's air up there? Our Moon is very different from Earth because it doesn't really have an atmosphere. It's almost completely in a **vacuum**! That's why when astronauts went to visit the Moon in the 1960s they had to wear space suits. I'm going to show you how to make your own moon-like environment by removing air molecules from a bottle. This one looks like a magic trick if you don't know what's going on. You're going to blow up a balloon inside a bottle using the power of a vacuum. You can even challenge your friends to put a balloon inside a bottle and see if they can blow it up. It's impossible! But I'm going to show YOU exactly how to do it—with science!

MAKE A HYPOTHESIS → What will happen when you remove your finger from the hole on the side? →

Materials

↓

Balloon

Plastic bottle

(you'll need one
that is pretty sturdy,
like a juice or
Gatorade bottle)

Thumbtack

INSTRUCTIONS

1. Place the balloon inside your bottle, but hold on to the opening of the balloon.
2. Fold the opening of the balloon along the lip of your plastic bottle.
3. Have your lab assistant (an adult!) use the thumbtack to poke a hole on the side of your bottle about three-quarters of the way down.
4. Place your mouth over the hole that you created and suck as much air as possible out of the bottle.
5. Without moving your mouth from the hole, very quickly place your finger over the hole (be careful not to let any air inside the bottle).

With every breath (where you are sucking in air from the bottle), you should see your balloon get bigger and bigger. At some point it will get hard to suck any more air out of the bottle.

"THUMBTACK HERE"

THE SCIENCE

Oh my gosh! The balloon grew bigger and bigger each time I sucked out more air!

Exactly right! You decreased the pressure inside the bottle. Air pressure can be decreased here in two ways. You can either remove air molecules or you can make the air colder. By sucking air out, you removed air molecules. So now, the air pressure outside of the bottle (in the room you are standing in right now) is greater than the air inside the bottle. That air pushed on the inside of the balloon, which made it take up more space inside your bottle.

Levitating Ring

Why do I shock my friends after going down the slide at the playground?

ABOUT THE EXPERIMENT

When you go down a plastic slide on the playground, electrons are rubbed off of the slide and stick onto you. Because electrons are negatively charged, you build up something known as a **static charge** on your body. This means that you have too many extra electrons sitting right on you, waiting to jump to something else. When you come across a friend, tap them with your finger and you'll feel a SHOCK! That shock is the feeling you get when electrons are jumping off of your body onto something (or somebody) else. With this experiment we're going to use static charge—not to shock anybody, but to make a ring levitate!

Materials
↓

Scissors

Produce bag
(the kind you get at the grocery store to put fruits and vegetables in)

Balloon

Cotton towel/ blanket
(or you could just use your hair)

88

INSTRUCTIONS

1. With scissors, cut off a ¼-inch (~½-cm) strip of your produce bag (unfold it so you end up with a plastic ring).
2. Blow up your balloon and tie it off.
3. Use your cotton towel to rub both the plastic ring and the balloon (or just rub the plastic ring and then the balloon against your hair—this is the way I like to do it).
4. Gently place the ring horizontally above the balloon and use the balloon to levitate the plastic ring.

You should be able to make your ring float above the balloon for at least 5 to 10 seconds. Be sure to keep the balloon under your ring because that is the thing that is pushing your ring up into the air.

TRY THIS
↓

Try different bags. You'll find that the repelling force from the static charge is only strong enough to lift really light bags. What if you only "charged" the balloon and not the bag? Will it work? Why or why not? You'll find that when you only charge one of the objects, you don't get that same propelling force because only one of them holds a negative static charge.

THE SCIENCE

I'm a magician! How am I doing this?

You're a scientist! When you rubbed the balloon against your hair, you

It's trying to fall down, but I won't let it!

You're right! Gravity is trying to bring it down, but luckily the bag is

Floating Stick Woman

How do you get permanent marker off a whiteboard?

ABOUT THE EXPERIMENT

Has your teacher ever warned you not to use permanent marker on their whiteboard? That's because permanent marker won't wipe off like a dry erase marker. But don't fear! There's a way you can remove it—with science!

This may sound silly, but you'll need to scribble all over your permanent marker with a dry erase marker. Then go ahead and wipe off everything with a dry eraser. The dry erase marker acts as a **solvent** and will help dissolve and lift the ink from your permanent marker, making it easier to wipe away.

Permanent markers are made up of acrylic **polymers**, which make them stick well to surfaces. But dry erase markers are made of oily silicone polymers, which make the ink stick well to itself but not flat surfaces in the same way permanent ink does. This allows us to erase dry erase markers from a whiteboard without any

MAKE A HYPOTHESIS → What would happen if you put dry erase ink in a cup of water and permanent marker in another cup of water? →

Materials
↓

Glass plate

Dry erase marker

Small container of water

(a liquid measuring cup would work, or even a plastic syringe—you just need to be able to pour it gently and slowly on top of your plate)

INSTRUCTIONS

1. **Draw a stick woman on your glass plate with the dry erase marker—make sure that there are no gaps in your stick figure, as all of her limbs need to be touching.**

2. **Slowly add droplets of water around your stick figure until you see it lift up from the surface and float.**

You should see your stick figure completely rise up on top of the water. If you move your plate around, the stick figure should move too.

water. In this experiment, we'll show off how cool dry erase markers really are by making a stick woman float on water (and even dance on your plate!).

THE SCIENCE

 Whoa! The stick woman rose off the plate and is floating on top of the water. But HOW?!

 Dry erase marker ink is really cool, huh? So, there are a couple of things going on here. One, dry erase marker is made up of chemicals that make it stick to each other—but it doesn't stick well to other things. This is how your stick figure could stay together, but it peeled up from the plate really easily.

 And it just rose up to the top of the water, too!

 Exactly! Dry erase ink doesn't dissolve in water. It's also less dense than water, so it floats right up to the surface! →

TRY THIS →

Draw a bunch of stick figures on your plate and slowly add water to them so that they all float along the surface. Then add soap to your finger and place your soapy finger in the center of your plate. Because soap is a **surfactant**, it will reduce the **surface tension** (weakening the bonds between water molecules) in the center of your plate and send your stick figures flying to the edges (where the bonds between water molecules are still strong).

NOTES!

Rising Candle

How does a candle work?

ABOUT THE EXPERIMENT

Next time your family lights a candle in your home, take a close look at it. How does it work? Where is that awesome smell coming from? Why can you sometimes see black smoke? Well, in a candle you have wax, a wick (that thing in the middle that you light), and a small flame. The flame uses the oxygen in the air and burns the wax through the process of **combustion**. This releases perfume into the air, which makes your home smell nice. Sometimes the flame doesn't get enough oxygen to properly burn all of the wax and you get some carbon left over—this is the black smoke that you see coming off of your candle. With this experiment, we're going to put some candles to work. In fact, we're going to make a candle "magically" levitate upward . . . using science! →

I NEED MORE OXYGEN!

Materials
↓

Water

Pie pan

Food coloring

Spoon

Tea-light candles

(many candles will work, but I think it looks cooler when the candle floats on top of water)

Lighter

Glass jar or vase

Safety note:
To do this experiment, you'll need to light something on fire, which requires adult supervision.

INSTRUCTIONS

1. Add water to your pie pan so that it fills about 1 inch (~2 cm) high.

2. Add food coloring to your water and mix it with a spoon.

3. Add your tea-light candles to the water (the number of candles will depend on the size of the opening of your jar—you want them all to fit inside).

4. Have an adult light your candles.

5. Place your jar or vase upside down on top of the candles.

Once you place your jar on top, you should see a few bubbles escape out of the bottom. Within a couple of seconds the flame(s) should go out and your water should start to rise up inside the jar (the height will depend on how many candles you have and the size of your jar).

MAKE A HYPOTHESIS
→ **What would happen if you added more candles?** → **Fewer?** →

THE SCIENCE

 Wow! The water rose up inside the glass jar, like magic!

 Like science! Did you notice anything else? What happened first?

 Yes! Well, first I saw bubbles right outside the glass vase.

 Exactly! The lit candles inside were heating up the air inside the jar. When air is heated up, the molecules move around faster and want to take up more space. Because of this, some of the air escaped. That's what those bubbles were. What did you notice next?

 So cool! Next, I noticed that the flames on the candles went out.

 That's right. Fire needs oxygen to survive. Once the candles used up all the oxygen in the jar, they went out.

 And then the water started rising inside the jar! Why did that happen?

 Two reasons. One is a chemical reason, and the other is a physical one. First, through the act of combustion—the candle flame—the paraffin wax and oxygen combine to create carbon dioxide and water. But the important part here is that it's burning more oxygen than it's creating carbon dioxide and water. So, the end result is that you have LESS air inside the jar.

 What was the physical reason?

 Well, you saw that when the air was heated up, some of that air escaped as bubbles, right?

 Because when air gets hot it wants to take up MORE space.

 Right! But then the candle went out. So, the air that was left inside started cooling down. When air cools down, it wants to take up LESS space—creating a lower pressure environment.

 Kind of like a vacuum ?

 Yes, kind of like a vacuum! And because the air pressure outside the jar is much higher, it pushes the water inside the jar, causing it to rise up.

Make your own guess, then flip upside down to read!

If you are able to add more candles, you'll be able to make the reaction even bigger. You'll heat up even more of the air and your water level should rise just slightly higher than before.

KITC
SCIE

HENCE

Extracting Star Poison from Cereal

How is iron created?

ABOUT THE EXPERIMENT

Iron is first created in the center of a massive star. Stars are like big, bright, beautiful engines that turn lighter elements into heavier ones. In really big stars, hydrogen is fused to create helium, which is fused to create carbon—and then oxygen, and then silicon, and then eventually silicon is fused to create iron. And that right there is the start of the star's death! Iron is like a poison to stars, because it costs so much energy to create iron that eventually the star will die—by exploding into a **supernova**, collapsing into a **neutron star**, or becoming a **black hole**.

In this experiment, we'll extract this star poison—ahem, I mean iron—from your cereal

using a magnet! I love this one because I think a lot of people are surprised to find this metal in things we eat. But don't worry, it's a good thing it's in there. Our bodies need iron to survive.

Materials

↓

2 cups (80 g) cereal with "reduced iron" in the ingredients list

1 cup (240 ml) water

Resealable gallon-size (4-qt) bag

Strong magnet

INSTRUCTIONS

1. Place the cereal and water in the bag.
2. Wait 1 hour.
3. Use the magnet to massage the bag for a minute or two (sometimes it's easiest to move the magnet toward the corner of the bag so you can gather the iron there).

After a couple minutes of massaging your bag, you should see tiny pieces of black metal floating around near your magnet. These will be really small (smaller than an ant!), so keep your eyes peeled. The more cereal you have, the more iron you will find.

TRY THIS

↓

Try moving your magnet around the bag. Can you make the little iron pellets move with it? Like little ants running toward your magnet? Try adding more cereal and see how much iron you can collect.

THE SCIENCE

 It looks like little ants marching around! How did that happen?

 Well, iron is attracted to your magnet. You can't see the little bits of iron when they are all spread out throughout your cereal, but you can bring them all together with the magnet.

 Wuahhaha! I can extract star poison from my cereal! Beware, my stellar friends! →

99

Wang Zhenyi was a famous Chinese scientist from the 1700s. During a time when women weren't allowed to study, she taught herself astronomy, mathematics, geography, and medicine. She even used everyday objects like a table, mirror, and lamp to research how a lunar eclipse worked!

NOTES!

_____

_____

_____

_____

_____

_____

_____

_____

DID YOU KNOW? →

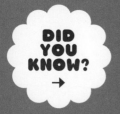

Our bodies need iron to survive. If you see "fortified with iron" on your cereal box, you may be thinking . . . like the METAL? And yes! It's in a slightly different form, but we do have iron in our cereal (and a bunch of other things we consume). Our bodies need iron to make hemoglobin, a protein in red blood cells that carries oxygen from the lungs to all parts of the body. One way to get that iron is by eating fortified cereal!

Cookie Science

Why is baking soda used in baked goods?

ABOUT THE EXPERIMENT

Have you ever wondered why we put baking soda or baking powder in baked goods? Those ingredients certainly don't taste very good on their own, so why would we include them in something as delicious as a cookie or a cake? Well, it has to do with science! If you've ever mixed baking soda with vinegar, you've seen that it creates an acid-base chemical reaction that makes carbon dioxide bubbles. This experiment will show you how to use that same science to make the look and texture of your cookies just right! You'll even see what cookies look like if you forget to add baking soda to the recipe. Roll up your sleeves, my little science chefs. This is a great one to do when you're hungry! And don't worry, the cookies with and without baking soda are both delicious. I tested them myself!

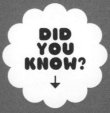

DID YOU KNOW?
↓

There is a ton of science in baking! Recipes are often very precise, so it's important to follow instructions just as you would follow steps in a scientific experiment. Different ingredients and steps (like mixing, folding, cooking, and frying) help the baker achieve just the right taste, texture, temperature, and presentation. A good chef is like a trained scientist in their kitchen!

→

Materials

Cookies with baking soda:

Dry Ingredients

1 cup plus 2 tablespoons (150 g) flour

½ teaspoon baking soda

¼ teaspoon salt

Wet Ingredients

½ cup (115 g) butter, softened

6 tablespoons (75 g) brown sugar

6 tablespoons (75 g) granulated sugar

½ egg

(you can mix one egg in a bowl and use half for this batch and half for the other)

½ teaspoon vanilla extract

1 cup (180 g) chocolate chips

Cookies without baking soda:

Dry Ingredients

1 cup plus 2 tablespoons (150 g) flour

¼ teaspoon salt

Wet Ingredients

½ cup butter, softened

6 tablespoons (75 g) brown sugar

6 tablespoons (75 g) granulated sugar

½ egg

(you can mix one egg in a bowl and use half for this batch and half for the other)

½ teaspoon vanilla extract

1 cup (180 g) chocolate chips

INSTRUCTIONS

1. Preheat your oven to 375 degrees Fahrenheit (190°C). Line two baking sheets with parchment paper.
2. You'll be making two batches of cookies. Do these steps for each batch.

3. Mix all of your dry ingredients in one bowl.
4. Mix all of your wet ingredients in another bowl.
5. Slowly add the dry ingredients to the wet ingredients and stir to combine.
6. Stir in the chocolate chips.
7. Scoop the dough into little balls and place on a prepared baking sheet, spacing them 2 inches (~5 cm) apart.
8. Bake for 8 to 10 minutes, or until golden and set around the edges.

You'll notice that the cookies without baking soda look very similar to the shape that they were when you first placed them in the oven. The cookies with baking soda should be spread out a bit more (with less space between the cookies than when you first placed them in the oven).

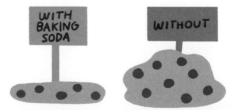

THE SCIENCE

 Huh, the cookies without baking soda are the same shape as when we put them in the oven. They kinda just look like a . . . ball. What happened?

 Baking soda is a leavening agent in cookies—this means it helps the dough rise. How does it do that? Well, baking soda is basic, which means it has a pH higher than 7. It reacts with ingredients that are acidic, or things that have a pH lower than 7, such as brown sugar, to create carbon dioxide bubbles. (See the pH scale in the back of the book for more information!) So, when you're mixing all of your cookie ingredients together, you are kick-starting an acid-base **chemical reaction** that creates little air bubbles in your dough. Those bubbles help stretch out the dough as it bakes, allowing the cookies to grow and spread out on your baking sheet.

 Ah, so the cookies without baking soda didn't have those bubbles, so they didn't stretch and grow. They just sorta stayed the same.

 Exactly!

 They're still delicious, though!

NOTES!

_____

_____

_____

_____

_____

Runaway Pepper

Why can some insects stand on water?

ABOUT THE EXPERIMENT

This experiment will show you the power of something called **surface tension**! Surface tension is the filmlike layer on top of a liquid that is created because water molecules are clingy. Literally! They are all slightly attracted to each other. It's kind of like they all really want to hold hands. When you have a glass of water, the water molecules in the middle are holding on tightly to the ones above them, below them, to their sides, and diagonally. But look at the top of the glass. Those molecules only have ones right below them and to their sides to hold on to. Because of this, they hold on extra tightly to the ones to their sides and the ones below them. This creates a little film on top of water that allows really light insects to literally walk on the surface.

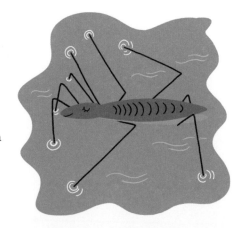

This experiment will show you how to break those bonds between water molecules using something known as a **surfactant**. It's very simple, easy, and fun for all ages! If you want to make it more colorful and sparkly, you can replace the pepper with really fine glitter (it just needs to be able to float on top of water).

Materials
↓

Plate
(best if it's the type of plate where the edges are raised a little so it can hold water)

Water

Pepper

Dish soap

INSTRUCTIONS

1. Fill your plate with water so that the entire plate is covered.
2. Sprinkle pepper all over the top.
3. Add dish soap to one finger.
4. Place your finger in the middle of the plate.

As soon as you place your soapy finger in the center of the plate, you should see the pepper rush away to the edges.

THE SCIENCE

 Whoa! The pepper ran to the edges! But . . . how?!

 The soap! Soap is something known as a surfactant. That means it makes the bonds between water molecules weaker and reduces the surface tension in the liquid. Remember how we talked about water molecules holding hands with each other? Well, soap is like a karate chop to that hand holding! It breaks the bonds between those molecules and takes away the film on top of water. So when you touch your finger to the center, the bonds between the water molecules are weaker there than the bonds between the molecules around the edge of the plate. Those stronger bonds pull at the water molecules in the center.

 And my pepper just rides the wave?

 Exactly! Your pepper is floating on top of the water and it's just riding the wave toward the edge.

WHO TO KNOW
↓

Maria Sibylla Merian was trained as an artist but was arguably one of the first true field ecologists. She used her artistic abilities to record detailed scientific observations of the life cycle of insects like caterpillars. By combining her love of art and science, she made historic contributions to the field of insect biology.

Soap Monster

When I'm in the pool, why do some toys float and some sink?

Lab Assistant (an Adult!) Required

ABOUT THE EXPERIMENT

hen you're playing in the pool, you'll notice that some toys will float on top of the water while others fall to the bottom. This is all about density. Pool noodles have a much lower density than water, which is why they float so well. Toys that are denser than water will sink to the bottom! Humans are just slightly denser than water. This is why we feel buoyant under water, but eventually we will sink to the bottom—especially if you let all the air out of your lungs, which would make your body even denser!

In this experiment, we'll play with a very special type of soap known as Ivory soap. This soap is special because it floats! Ivory soap has lots of teeny tiny air bubbles in it that make this type of soap less dense than water. Because of this low density, if you dropped it in the bathtub, it would float right to the top. Now . . . let's see what happens when you put it in the microwave!

Materials
↓

Bowl of water

Ivory soap

Microwave-safe plate

INSTRUCTIONS

1. Take out your Ivory soap and place it in your bowl of water. Note what happens!
2. Now, take your Ivory soap and place it on a microwave-safe plate.
3. Have your lab assistant place that plate in the microwave for 90 seconds.

You'll notice that the soap floats in water, just as advertised! Watch as the time passes in the microwave—the soap should be getting bigger and bigger in all different directions. It should look like it's coming alive—like a soap monster—right in your microwave!

TRY THIS
↓

Break off a piece of your microwaved Ivory soap and place it in the same bowl of water. You'll notice that it floats even higher (more of the soap is above the water) than the original soap did. When you microwaved the soap, you expanded each of those little air pockets inside, making the microwaved soap even less dense than it was before.

THE SCIENCE

Wow! That grew like a whipped cream monster taking over my microwave! Why'd it do that?

Well, Ivory soap is very special. It has a ton of little air bubbles all throughout the soap. This is what makes it feel so light and airy. When you heat it up in the microwave, you're heating up all of those little pockets of air. When air heats up, it wants to take up more space. Also, the soap is becoming warmer and a bit stretchier. So, each of those air pockets is growing and it's stretching the soap along with it. The result is a whipped, wavy, cloud-looking soap!

WE'RE HEATING UP

MAKE ROOM !!!

Self-Imploding Can

Why do astronauts need space suits?

ABOUT THE EXPERIMENT

 *I always see astronauts wearing these big, bulky space suits when they are doing space walks. Why can't they just wear their regular clothes? I bet it would be more comfortable!*

You're right, it would be more comfortable! But those space suits are very important. They are providing air **pressure** on astronauts' bodies.

 Air pressure? Why do they need air pressure?

You may not realize it, but the air in the room you're standing in is bearing down on your body and everything around you. It's as heavy as a milk jug on the tip of your finger!

 And that's a good thing? I mean, it seems like it would crush us. That's so heavy!

 I'm going to blow your mind. Not only does it not crush us, but we need it to survive! Our bodies evolved to require this level of air pressure to keep all the pockets of air inside our body, like in our lungs, where they are! In fact, when we go to the **vacuum** of space, where there is no air, we need to bring our own pressure with us—that's one of the reasons humans have to wear space suits in space—to push back on our bodies to provide the pressure we're used to here on the ground.

Materials
↓

¹⁄₃ cup (80 ml) water

Empty soda can

Stove top

Ice bath
(a big bowl of ice and water)

Tongs

This experiment will use that intense air pressure in the room and show you just how powerful it is. This one always makes me jump!

INSTRUCTIONS

1. Pour the water into your soda can.

2. Grab your lab assistant (an adult) and have them set your stove top on high heat.

3. Have your lab assistant place your soda can on the stove top and wait 5 minutes.

4. Have your ice bath ready.

5. After 5 minutes, have your lab assistant use the tongs to grab the soda can quickly and then carefully flip it upside down into the ice bath.

6. You need to do this really quickly! If it doesn't work the first time, just start back at step 1 and try it again.

The can should implode instantly! If it doesn't happen within 1 second of moving to your ice bath, you should try again from step 1. →

THE SCIENCE

Wow, it imploded in the bowl! Why did that happen?

You are witnessing the power of air! When you boiled the water in the can, you turned the water into water vapor that rose and pushed most of the air molecules out of the can.

Poor air!

Don't worry, air molecules don't have feelings! So now, you have a can filled with water vapor and barely any air molecules. Your assistant flipped that over into the cold ice water. The cold environment instantly condensed the water vapor back into water, which takes up a lot less space in the can. Because there weren't many air molecules left in the can, this created a little bit of a vacuum in the rest of that can. So, you have a super low-pressure environment inside the can and a relatively high-pressure environment outside the can. Because of this difference, the air pressure in the room was able to crush your can completely!

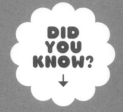
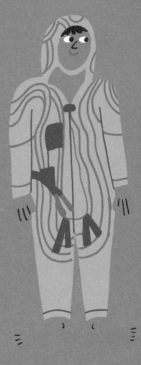

NOTES!

THIS MAY GET MESSY

Oobleck

Why is ketchup so hard to get out of the bottle?

ABOUT THE EXPERIMENT

Shake, shake, shake! Ever wonder why you need to shake and hit a ketchup bottle to get the ketchup out? Ketchup is a special type of fluid known as a **non-Newtonian fluid**. That means it doesn't abide by normal fluid rules. Sometimes it acts like a liquid (when force is applied) and like a solid (when no force is applied). When you shake and hit the bottle, the ketchup becomes more liquidy and can escape.

Oobleck is another non-Newtonian fluid. But it's the opposite of ketchup—when you add force it acts like a solid, and when you don't add force it acts like a liquid. I used to make Oobleck as a kid with my mom, and it was always a dream of mine to fill an entire pool with Oobleck! So, when we were able to do that on *Emily's Wonder Lab*, it was a dream come true! I hope you love it as much as I do.

Materials
↓

1 cup (140 g) cornstarch

½ to ¾ cup (120 to 180 ml) water

2 drops of food coloring
(optional)

Bowl

MAKE A HYPOTHESIS → If we made an entire pool filled with Oobleck, what would you need to do to get across it without sinking—walk? Crawl? Hop? Dance? Run? →

Warning:
Oobleck might clog your drain, so when you're done, throw it in the trash.

INSTRUCTIONS

This recipe isn't exact. If it feels too dry, add a little water. If it feels too watery, add a little cornstarch. And remember, you can make as little or as much as you like—you can fill an entire pool with this stuff, in fact!

1. **Add the cornstarch, water, and food coloring (if using) to a bowl.**

2. **Mix with your hands (this will get messy!).**

This may look a little liquidy at first, but try to squeeze it into a ball—does it feel like a solid? If not, add a little bit more cornstarch. If it doesn't feel liquidy at all, add a little bit of water. You should be able to pick it up, squeeze it into a ball, and then stop squeezing and let it fall through your fingers.

THE SCIENCE

 Whoa, this feels so weird! Sometimes it's hard, but sometimes it feels like a liquid.

 That's what makes Oobleck a non-Newtonian fluid. You can change what it feels like by adding or removing **pressure.** *Squeeze it together and it feels like a solid, but if you stop squeezing it, it goes back to a liquid!* →

(Make your own guess, then flip upside down to read!)

When we tried this on Emily's Wonder Lab, the kids who made it across the pool of Oobleck without sinking were the ones who applied a force to the Oobleck: like hopping, running, or even dancing. Basically, they pushed on the Oobleck in some way. Remember, Oobleck acts like a solid when force is applied—that's the key to getting across!

Grab a chunk of your Oobleck and pass it between your hands like a ball. It stays together because you are adding force to your Oobleck, making it feel like a solid. Now, go ahead and stop passing the ball and just hold it in your hand. It falls through your fingers like a liquid. That's because when you stop adding force it acts like a liquid. Now try to let your hand sink to the bottom of your bowl and then lift your hand really fast. Did you bring up the bowl with you? That's because the force of your hand lifting kept the Oobleck solid, and it held on to your hand like a monster from the deep!

NOTES!

Magic Unicorn Bubbles

Why are baking soda and vinegar so good at cleaning stuff?

ABOUT THE EXPERIMENT

This experiment is a great one to do outside—but if you want to clean your bathtub, you can do it in there too. Baking soda is a base and can help dissolve dirt and grease. Vinegar is an acid that can break up water stains on the tub. (To learn more about acids, check out the pH scale in the back of the book!) When you mix the two together, it creates bubbles that can help lift and move the dirt to the drain! But be warned, this experiment can get very messy, especially with little ones! To make these "unicorn bubbles," I recommend using pink, blue, and purple food coloring, each in different tins. →

Materials
↓

Food coloring in multiple colors

(liquid food coloring works best, but gel-based food coloring will also work)

Muffin tin

(or a bowl or baking pan)

1 cup (230 g) baking soda

(you may have some left over)

2 cups (480 ml) vinegar

(you may have some left over)

Small cup

Spoon

INSTRUCTIONS

1. Place 5 drops of food coloring in each muffin tin.

2. Place 1 spoonful of baking soda in each tin.

4. Use a small cup to pour about 2 spoon-fuls of vinegar in each tin, little by little.

3. Shake the tin a little to even out the baking soda and cover the food coloring.

NOTES!

_____

_____

_____

_____

Did you see that? You probably didn't even need all of your vinegar to get a ton of bubbles! You are a chemist who has created an acid-base **chemical reaction** and carbon dioxide bubbles. Now try to guess what the next color will be—do you remember? Go ahead and pour 2 spoonfuls of vinegar in the next tin.

5. **Keep repeating this until you've revealed all the colors!**

As soon as you add the vinegar you should see bubbles form. The color of those bubbles should be the color of the food coloring you added.

THE SCIENCE

 You can't see the colors at first because the baking soda is a solid, which means its molecules aren't moving around very much. So, the food coloring is just sitting there, under the baking soda, waiting to be revealed. All it needs is a liquid.

 And that's where the vinegar comes in!

 Precisely! The vinegar makes two things happen. As a liquid, it's making the molecules in the tin dance around so now the food coloring can mix together with the solution. And secondly, vinegar is an acid—its main ingredient is known as acetic acid—that you are adding to your base, baking soda, also known as sodium bicarbonate. By mixing these two ingredients together, you are starting a chemical reaction that creates carbon dioxide (that's the bubbles you see!), water, and a type of salt called sodium acetate.

Floating Water

How heavy is . . . air?

ABOUT THE EXPERIMENT

Have you ever wondered if air has weight? It seems so light and . . . airy, doesn't it? But it's actually quite heavy! In fact, the air in the room around you right now is bearing down on all of your muscles and bones. At sea level, air puts 14 pounds per square inch of pressure on you! That's as heavy as a jug of milk on the tip of your finger.

In this experiment, we'll learn how air pressure can be used to hold a glass of water upside down. This one is great to do outside or over a sink.

MAKE A HYPOTHESIS → What would happen if you tilted your glass? → Why? →

Materials

↓

Glass of water filled three-quarters full

(make sure the rim of the glass is smooth and level)

Small square of cardboard or an index card

(this just needs to be large enough to cover your glass's opening)

INSTRUCTIONS

1. **Bring your glass of water either over your sink or outside.**
2. **Place your cardboard on top of your glass.**
3. **Hold your hand on the top to keep the cardboard closed over the opening.**
4. **Carefully flip your glass upside down with the cardboard sealed tightly over the opening.**
5. **When your cup is straight and level, remove the hand holding the cardboard.**

The water should stay inside your glass. Keep the cup level and don't move it around too much, or else the water will spill out.

Step 2

Step 5

THE SCIENCE

 Whoa! The water stayed in my glass! How is that happening?!

 It's because of the air pressure in the room! Think of air pressure as little fists that are punching you and everything around you. Those little air fists are punching on the glass everywhere, including the bottom of the glass where the cardboard is. The force of the atmospheric air on

your cardboard is greater than the combined force of the weight of water bearing down on your cardboard and the little amount of air still left in your glass. The cohesive forces of surface tension help keep the cardboard in place, but it's the incredibly strong air pressure from the air around you that's preventing the water from pushing the cardboard down and spilling out everywhere.

(Make your own guess, then flip upside down to read!)

When you tilt the cup, more water is concentrated on one side of the cardboard, pushing harder on one side of the cardboard than the other. Now, the surface tension isn't strong enough to hold the cardboard in place. The cardboard falls off and the water comes out with it.

Unicorn Slime

How does glue work?

ABOUT THE EXPERIMENT

L et's say you want to glue two pieces of paper together. You can't see it with your eyes, but if you looked at paper under a microscope it would look like a bunch of little hairs overlapping each other, with a bunch of little gaps between them. When you put glue on paper, because it's so liquidy, it can flow into those gaps on both pieces of paper—kind of like a claw snaking its fingers around all those hairs. The magic happens once the glue dries. After it's dried, the glue becomes a solid—those claws become stuck in place with their fingers holding on to all the little hairs it flowed around. So, the glue is grabbing both pieces of paper at once! That's how the papers stay stuck together.

In this experiment we'll learn why glue is so runny and how to transform it into something a bit more . . . slimy!

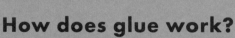

Materials
↓

⅓ cup (80 ml) glue
(make sure your glue contains PVA)

2 tablespoons water

¼ teaspoon baking soda

Bowl

Spoon

1 cup (~240 ml) shaving cream

Food coloring
(whatever color you like)

¾ tablespoon eye contact solution
(must contain boric acid and sodium borate)

INSTRUCTIONS

1. Pour the glue, water, and baking soda into a bowl and stir with a spoon.
2. Add the shaving cream and stir again.
3. Add a couple drops of food coloring and stir.
4. Add the contact solution and stir.

You should see the slime start to form as you mix in your contact solution! Be sure to fully mix all of your ingredients.

TRY THIS
↓

Try making a second unicorn slime batch, but this time in a different color. Now you can swirl them together and make it look like slime meant for the most beautiful unicorn!

THE SCIENCE

 Fluffy slime is my favorite slime! How did it get so . . . fluffy?

 Well, when you poured the glue into the bowl, what did you notice?

 I noticed that the glue was really flowy and kind of liquidy.

 That's right! That's because glue is something known as a **polymer**. A polymer is something that has long, repeating chains of molecules in it. Right now, in the glue, those long repeating chains of molecules can flow past each other really easily. Kind of like rowboats down a big river. But what happened after you added the contact solution?

 It immediately felt weird when I started stirring it! The foamy glue started sticking together a bit more and it slowly became slimier.

 Yes! That's because the baking soda and the contact solution acted as an activator. An activator is something that all slime recipes have. The activator is very important because it cross-links those long repeating chains of molecules we talked about.

 The rowboats?

 Yes, the activator is kind of like a rope that ties all the rowboats in a river together, so they no longer flow past each other as easily. And that's what makes it less liquidy and more slimy!

 So, what does the shaving cream do?

 It just makes it fluffy! Because everyone knows unicorn slime is the fluffiest!

Horse Toothpaste

How do bubbles work?

ABOUT THE EXPERIMENT

Have you ever tried to blow a bubble with just water (and no soap)? It doesn't work! That's because the **surface tension** in water is too high. This means that the water molecules are holding on to each other so tightly that the water isn't flexible enough to form a bubble. This is where the magic of soap comes in! Soap reduces the surface tension in water (it wiggles itself in between the molecules and makes those bonds between water molecules weaker)— and this makes the liquid more flexible. So, we have soap to thank for our ability to trap air into those pretty little bubbles.

In this experiment, we're going to trap a *lot* of air using soap and create a color-ful, foamy mess! This was one of the most

popular experiments on *Emily's Wonder Lab* because it was so colorful and explosive. The kids loved it, I loved it, and the horses loved it (they watched from afar)! I'm going to show you a smaller version of horse tooth-paste that you can do safely at home.

Materials
↓

Safety glasses

Plastic gloves

Potassium iodide crystals

(you can buy this online)

Glass with a lip to pour

(like a liquid measuring cup)

½ cup (120 ml) water

Spoon

Container

(I recommend a chemistry flask, but any type of container that is large on the bottom and smaller up top will work—like a bottle, vase, etc.)

Hydrogen peroxide 3% concentration

(you can buy this at your local pharmacy)

Food coloring

Dish soap

INSTRUCTIONS

1. Put on your safety glasses and gloves.
2. Place 1 tablespoon of potassium iodide crystals in your liquid measuring cup.
3. Add the water. Use a spoon to mix it up.
4. Place your flask (or vase) on a flat surface (I recommend doing this outside).
5. Pour ½ cup (120 ml) of the hydrogen peroxide into your flask.
6. Place 5 drops of food coloring in your flask.
7. Squirt about 1 spoonful of dish soap in your flask. Carefully swirl it around.
8. Now, take your liquid measuring cup that has the potassium iodide in it . . . and dump it in your flask in 3 . . . 2 . . . 1!

Safety note:

Make sure you don't eat any of these chemicals, and always wear protective goggles and gloves so you don't get chemicals in your eyes. It's okay if you get this stuff on your skin, we just want to make sure you don't get it in your mouth or eyes. We only call it "horse toothpaste" because it looks like toothpaste. Don't put this in your mouth! You'll need an adult to help you with this one.

→

DID YOU KNOW?
↓

You might have noticed that the horse tooth-paste we created on *Emily's Wonder Lab* shot a lot higher into the air. That's because we used a more concentrated form of hydrogen peroxide. This type of hydrogen peroxide is only for professionals and you can only buy it from chemical supply companies. But if you want your reaction to be a little more dramatic, put it in a container that has a large bottom and a small opening (like the shape of a volcano). You'll be forcing the larger reaction that's happening at the base out of the smaller opening at the top, which will make it move out faster!

THE SCIENCE

 Whoa, it's like mini horse toothpaste erupting out of my container!

 That's right! You're witnessing a chemical reaction . . . in action! The chemical name for hydrogen peroxide is H_2O_2—basically like water, but with an extra oxygen molecule. When we add the potassium iodide, it removes that extra oxygen molecule, so all you have left is water and O_2 (oxygen gas)!

 What makes it so foamy?

 That's why we add the dish soap! The soap traps the oxygen gas into soapy bubbles, which makes the reaction so toothpaste-like.

 Let's do that again!

Yes! This time, let's try different colors to find which one is our favorite!

NOTES!

_____

_____

_____

_____

_____

Surface Tension Extravaganza

Are raindrops actually shaped like teardrops?

ABOUT THE EXPERIMENT

Have you ever wondered whether a raindrop falling from the sky actually looks like the shape of a teardrop? It doesn't! Really small raindrops look spherical—like a ball. This is because the bonds between water molecules are so strong (this creates a force known as **surface tension**) that they pull the shape of a raindrop together tightly into a sphere. As a raindrop gets larger, other forces—like the air hitting it as it falls—affect its shape too, so some larger raindrops take on funnier bean-like shapes.

In this experiment, we're putting those forces between water molecules to work in a couple of different surface tension experiments. Don't worry if it's a rainy day. These can be done indoors! →

Materials
↓

Plastic syringe
(I use the plastic syringe that comes with baby medicine)

Penny

Retractable pen

Glass of water

INSTRUCTIONS

Water on a Penny

1. Fill up your plastic syringe with water.
2. Set your penny on a flat surface.
3. Slowly add drops of water to the top of the penny.
4. See how many you can add before it spills over the side.

You should see water build up on top of your penny like a small dome.

THE SCIENCE

 The water looks like a little dome on top of the penny!

 That's right! It can build up really high because the water molecules on the surface are holding on to each other tightly!

 But eventually it spilled over. Why's that?

 Because even though surface tension is pretty strong, it can only hold on to the water to a certain point. If you add too much water, it'll get too heavy and spill over the side.

 Is surface tension the same reason why the spring is floating?

 That's exactly right! This is even more impressive because that spring is made of metal, which is denser than water. We know that it would normally sink right to the bottom. But because of surface tension, it's able to float on top. This is like those bugs that can walk on top of water!

MAKE A HYPOTHESIS
→ **What would happen to the spring if you gently touched the surface of the water with soap?** → (Make your own guess, then flip upside down to read!)

Floating Spring

1. Remove the top of your pen to reveal the spring inside.

2. Fill your glass of water to the top.

3. Hold the sides of the spring with the very tips of your fingers.

4. Slowly and carefully place the spring on top of the water.

You should see your spring float on top of your cup of water. You need a steady hand to do this one! If you are having trouble getting your spring to float, try asking your lab assistant (an adult) for help.

NOTES!

_____

_____

_____

_____

_____

By touching the water with soap, you are breaking the bonds between the water molecules, which means you are reducing the surface tension of the water. The surface tension was the reason your spring could float on top of water (even though it is denser than water), so as soon as you break the surface tension, that spring will fall into your cup!

SPINI
SCIEI

132
**Milk
Fireworks**

134
**Mini
Vortex
Blaster**

137
**Jumbo
Vortex
Blaster**

NING
NCE

140
Tornado
in a Bottle

Milk Fireworks

Why is soap so good at cleaning your hands?

ABOUT THE EXPERIMENT

Have you ever tried to wash oil off your hands without soap? Go ahead, try it! It's practically impossible. You need soap! Soap is made up of **surfactants**, which are itty-bitty chains of hydrogen, carbon, and oxygen atoms. On one end of the chain is the side that loves fat and oil (this is called **lipophilic**) and on the other end is the side that loves water (this is called **hydrophilic**). So, when you wash your hands with soap, one end of those chains is grabbing on to the oil and fat on your hands. And then when you rinse your hands, the other end of those chains grabs on to the water, falls off your hands, and takes all of that oil and fat down with it.

Soap is a very powerful tool that will allow us to create beautiful fireworks—in our milk! I've done this experiment a thousand times because it's so fun, easy, and downright gorgeous. I love it!

GO AHEAD, TRY IT !!!

Plate

Milk
(any type will work)

Food coloring
(water-based food coloring works best; if you only have the gel kind, just water it down a little first)

Cotton swab

Dish soap

INSTRUCTIONS

1. Fill your plate with milk so that it covers the whole plate.
2. Add drops of different food coloring to the center of the plate, so that they are all touching each other (or very, very close).
3. Dip one end of your cotton swab into dish soap.
4. Dip the soapy end of your cotton swab to the center of the plate where all of your food coloring is.

As soon as you add the cotton swab to your milky plate you should see food coloring jolt out in different directions toward the edge of the plate.

Step 2

Step 3

Step 4

THE SCIENCE

It's so pretty! What is making all of that food coloring move?

Two things are happening here. First, the soap is reducing the **surface tension** in the middle and the molecules on the top of the milk are getting pulled outward to where the surface tension is strongest. Also, remember how soap is made up of chains that have a fat-loving end and a water-loving end? Well, milk is made up of both fat and water. So, the soap molecules are dancing in the milk, and this is causing all of those beautiful swirls you see!

DID YOU KNOW?
↓

The color of real fireworks depends on the **chemical** elements that are burning. Strontium creates a red firework, barium makes green, copper creates blue, and sodium explodes into an orange/yellow color. Next time you watch a fireworks display, know that you are witnessing beautiful, explosive chemical reactions right before your eyes!

Mini Vortex Blaster

How is wind created?

ABOUT THE EXPERIMENT

 ind is created when you have an area of high **pressure** that meets an area of low pressure. This difference in pressure causes air to move and mix, which we feel as wind. In this experiment, I'll show you how to create your own wind in the form of a mini **vortex**. You'll need some way to create smoke or fog. If you choose to use a match, you'll need adult supervision. If you have a fog machine on hand, that would be even better!

Materials
↓

Scissors

Plastic bottle

Matches
(or fog machine)

Balloon

INSTRUCTIONS

1. With scissors, cut off the bottom 1 inch (~2 cm) or so from the bottle.

2. Cut the opening of your balloon off (about 1 inch [~2 cm] from the opening).

3. Place your balloon on the bottom of your bottle.

4. Have an adult light two or three matches, and then quickly blow them out and toss them in your bottle.

5. Hold your hand over the opening of the bottle so that the smoke can build up a bit.

6. Now gently tap the balloon end of your bottle to blast your vortex rings!

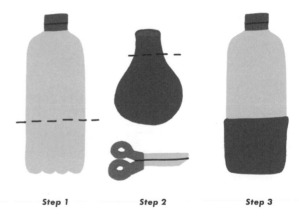

Step 1   Step 2   Step 3

You should see small donut-shaped rings fly out of your bottle. These may be hard to see. If you can't see them, try making them in front of a darker background (like a dark wall or blanket), or you may need more fog or smoke inside your bottle. →

WHO TO KNOW
↓

June Bacon-Bercey was the first African American woman to deliver the weather on television as a trained meteorologist. Her pioneering work helped make weather science more relatable to minorities and other women.

THE SCIENCE

 It's like I'm creating SMOKY DONUTS! Why do they have a hole in the middle?

 Well, when you tap the bottom of your bottle, you're pushing a ball of air through the opening. The air in the center of that opening goes the fastest, creating an area of low pressure (fast-moving air creates lower pressure), but the air around the perimeter goes the slowest (so this is an area of high pressure). Why does it go slower? Well, it's getting slowed down by friction because it's touching the inside of the bottle. So, you have an area of low pressure in the middle surrounded by an area of high pressure—that difference in pressure makes the air move and mix. We see this moving air as a vortex.

 And that vortex kind of looks like a donut!

 Yes, it does!

TRY THIS
↓

Try this with a friend—let's see how far your vortex rings will fly! First, try just blowing air on your friend when they are sitting about 1 foot (0.3 m) away . . . then 2 feet (0.6 m) away . . . then 3 feet (1 m) away. When did they stop feeling the air? Now let's try it with your vortex blaster. How much farther away can they sit and still feel the air? Your vortex blaster allows you to move a lot of air through a small opening, which makes it go much faster than the air you blow out of your mouth— this allows it to travel farther.

NOTES!

Jumbo Vortex Blaster

Does the Moon have weather?

Lab Assistant (an Adult!) Required

ABOUT THE EXPERIMENT

Ever wonder whether the Moon has storms, or hurricanes, or tornadoes, or even wind like we have here on Earth? Many other planets and moons in our solar system have storms, but not our Moon! Because of this, the craters that exist on its surface can stay there for millions of years! Our weather systems here on Earth (like wind and rain) along with our geological activity (like volcanoes and earthquakes) help recycle our planet. So, any crater that is formed here on our planet likely won't stay here very long. At least not for millions of years like they do on the Moon.

This experiment will show you how to create your own jumbo wind vortex—something big enough to make the Moon jealous! We made these on *Emily's Wonder Lab* and they were so much fun! This one involves cutting a trash can with power tools, so make sure you have your lab assistant (an adult!) handy. ➔

Materials
↓

Tool to cut into trash can

Large plastic trash can

Plastic shower curtain

Bungee cords

Scissors

Fog machine

INSTRUCTIONS

1. Use your cutting tool (I've used a utility knife, or a power tool with a cutting wheel attachment) to cut a 1-foot (0.3-m) diameter hole in the bottom of your trash can.

2. Flip your trash can upright.

3. Place your shower curtain over the large opening of the trash can and secure it with bungee cords.

4. Use your scissors to cut off the excess shower curtain.

5. Your jumbo vortex blaster is ready! Bring it to a location where there is not much wind.

6. Turn on your fog machine and fill your vortex blaster with fog (10 to 20 seconds).

7. With an open hand, tap the shower curtain end of your trash can to blast your jumbo-size vortex rings!

THE SCIENCE

 Whoa! How did we make those so much bigger?

 Well, we created a way to force more air outside a larger hole, so the end result was a larger vortex ring!

 How far will the vortex ring fly?

 Let's test it out. Take 5 steps back, now 10 steps, now 15 steps. Do you feel the blast?

 Yes, and it works even without the fog machine.

 That's right! The fog just helps us see the vortex. The vortex is always there.

You should see a large donut-shaped vortex blast out of your trash can and travel at least 10 feet (3 m) before breaking up. If you don't see anything, make sure it's not too windy where you are (the wind can break up the donut quickly), and that you have enough fog inside your vortex blaster.

NOTES!

_____

_____

_____

_____

_____

_____

TRY THIS
↓

Set up six plastic cups on top of each other like a pyramid (three on the bottom, two on top of those, and then one on the very top). Take 5 to 10 steps back and see if you can knock them down with your Jumbo Vortex Blaster. It's like bowling . . . but with air!

Tornado in a Bottle

How are tornadoes created?

ABOUT THE EXPERIMENT

Tornadoes form when warm, humid air (higher **pressure** air) collides with colder air (lower pressure air) to create a swirling **vortex**. On *Emily's Wonder Lab* we created our own tornado! The main ingredients for an indoor tornado are (1) spinning air and (2) air moving upward. We created the spinning air with five fans that we angled around in a circle. That made the air spin in a circle. Then we placed a really powerful fan on the ceiling and flipped it upside down, so instead of blowing air downward, it sucked air upward! Voilà! An indoor tornado!

In this experiment, I'll show you how to create your own tornado. But don't worry—this one won't take up the size of your room! It'll fit nicely inside a 2-liter bottle.

Materials
↓

2 empty
2-liter plastic
bottles

Water

Colorful
lamp oil

Vortex
connector
(you can buy this online)

INSTRUCTIONS

1. **Fill one of your bottles three-quarters full with water.**
2. **Fill that same bottle the rest of the way with your colorful lamp oil.**
3. **Connect your vortex connector.**
4. **Connect your other (empty) bottle to the vortex connector and twist it on tight.**
5. **Flip over your two-bottle system and then spin it to twirl the liquid inside.**

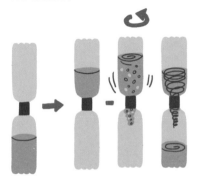

As soon as you stop spinning your bottle, you should see the liquid twirling down the connector and into the bottle below.

TRY THIS
↓

Fill a 2-liter bottle with water. Now, you're going to try to get the water out of it the fastest way you can (do this over a sink or outside)—what do you think is the fastest way to get the water out?

Try simply turning the bottle upside down and use a timer to figure out how long it takes for the bottle to empty.

Now, refill it, and this time, twirl the bottle so that you create a vortex, just like you did in the experiment. Time how long it takes to empty the bottle. It was faster the second time, right? That's because you created a vortex. The water swirls around the outside while the air rushes up through the middle. The first time you probably heard some "gulp, gulp, gulp" sounds, right? That's because the air coming into the bottle was getting in the way of the water going out. A vortex makes the whole process more efficient!

THE SCIENCE

 I made my own tornado!

 That's right! One of the main ingredients in a real tornado is spinning air. You created that same effect with water by spinning your bottle around.

 And the lamp oil makes it look really pretty too!

 It sure does! Lamp oil is less **dense** than water, so it will sit right on top of the water in your bottle. That density difference creates a really pretty effect when you swirl the liquids together.

FUN WITH PHYSICS

Egg Inertia

Why are car crashes so dangerous?

ABOUT THE EXPERIMENT

Car crashes are so dangerous because of Newton's First Law of Motion, sometimes called the **Law of Inertia** . This law tells us that an object in motion wants to stay in motion. If we are sitting in a car that is going really fast and it suddenly stops because it runs into something—our bodies still want to go at the same speed the car was going before it crashed. We don't want this to happen—this is why seat belts are so important to wear! They help keep our bodies in our seats when they really want to stay in motion! A seat belt helps stop your body from moving and can even save your life. There's another part to the law of inertia— an object in motion wants to stay in motion . . . but an object at rest wants to stay at rest! This experiment will show you how to use **Newton's First Law of Motion** — the Law of Inertia—to do a fun trick with an egg. It may take a couple of tries to get it just right, so be sure to try this one in an area that's okay to get messy, like outside.

Materials
↓

Glass of water

Tin pie pan

Toilet paper tube

1 raw egg

INSTRUCTIONS

1. Set your glass of water (filled about halfway) on a flat surface.

2. Place your tin pie pan on top, facing upward.

3. Place your toilet paper tube vertically on top of the pie pan.

4. Place your egg horizontally on top of the toilet paper tube.

5. Open your hand and keep your fingers together. Using the palm side of your hand, quickly knock the pie pan off the glass horizontally (being very careful not to hit the glass or anything else).

As you quickly knock the pie pan out of the way, your egg should drop directly into your cup! If this didn't work, you may need to hit the pie pan more quickly.

WHO TO KNOW ↓

Miriam Menkin was a geneticist and the first person to fertilize a human egg outside of the body. It's because of her work that families who have a hard time getting pregnant naturally have another method—something called in vitro fertilization or IVF—that they can use to help them make a baby!

THE SCIENCE

 WHAT! How did the egg end up in the glass?! Why didn't it fly away with the pie pan?

 Because an object at rest wants to stay at rest unless an outside force acts upon it. The egg was happily sitting right above your glass of water. When you hit the pie pan, that force made it fly sideways.

 And it looked like my toilet paper roll tumbled sideways too?

 Yes! That's because the toilet paper roll was resting on top of the pie pan. Friction between the bottom of the toilet paper roll and the top of the pie pan caused the bottom of the roll to slide sideways too. That's what made it tumble away.

 And then my egg just PLOPPED right down into my glass!

 Exactly! With the toilet paper roll swept out of the way, gravity took over and made the egg fall directly down into the glass.

145

Balancing Act with a Broomstick

Why are squirrels so good at balancing?

ABOUT THE EXPERIMENT

Squirrels may not be the smartest creatures in the animal kingdom, but they sure are good at balance! You may see them running across branches or even telephone wires without missing a beat. How do they do this? First, their tails can swish back and forth, changing something known as their **center of mass** , as they walk. Their tail can also act like a little parachute if they fall. Squirrels' rear ankles can also rotate nearly 180 degrees. These flexible ankles help squirrels shift their weight so they can balance in almost any scenario.

In this experiment, we'll learn more about the center of mass using just a broomstick and your hands.

INSTRUCTIONS

1. Have a friend hold the broomstick horizontally in front of you.

2. Hold out two fingers on each hand and balance the broomstick on the fingers, with your hands about 2 feet (0.6 m) apart.

3. Slowly inch your fingers together until both hands are touching, making sure you keep the broomstick balanced.

Where did your hands end up?

WHO TO KNOW
↓

Simone Biles is the first woman to successfully complete the Yurchenko double pike. This is a gymnastics move that involves a round-off onto a springboard, back handspring onto a vaulting table, and a double flip. It's something that is very, very difficult. Gymnastics, especially such difficult moves like this one, involve a lot of balance and physics. Simone Biles is a gymnast with unprecedented strength and control, which allows her to rotate faster when she is in the air while still maintaining her balance as she lands back on the ground.

THE SCIENCE

 This experiment is all about the center of mass. The center of mass of an object is the average position of all the mass in that object. To balance an object without it tipping, you need to place a balancing point under its center of mass.

 Ohhh . . . so my fingers are the balancing points?

 Exactly! Try balancing the broomstick with both of your hands at the exact center. What happens?

 It falls down!

 That's right. You moved your balancing points away from the center of mass.

 But why isn't the center of mass . . . in the center of the broomstick?

 If you look at the broomstick, you'll notice a lot of mass where the bristles are. That moves the center of mass closer to that part of the broom.

 And that's why I need to hold my hands closer to the bristles!

 Precisely!

Balancing Act with a Fork and Spoon

How does the balancing bird toy work?

ABOUT THE EXPERIMENT

Have you ever played with a balancing bird toy? Isn't it cool how you can hold the entire toy up by balancing the bird's beak on your finger? It looks like it's flying! But so much of the body is behind its head. Why doesn't the bird fall backward? This works because its wings are heavier than the rest of the body and they are both stretched out in front of its head, making the best place to balance the entire bird right under its nose. The heavy wings changed the location of the bird's center of mass. In this experiment, we'll create our own version of the balancing bird with a fork, a spoon, and a match.

Materials

Metal fork
(you'll need a fork that has prongs that can bend relatively easily and permission to do this)

Metal spoon

Match

Drinking glass

Lighter

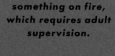

Safety note:
To do this experiment, you'll need to light something on fire, which requires adult supervision.

INSTRUCTIONS

1. Bend the middle two prongs of the fork slightly upward so that you can fit the round part of the spoon inside (you want them to be able to stay together easily and not fall apart). The shape will look a bit like a "C."

2. Place a match in between the middle prongs of the fork.

3. Balance your fork-spoon-match system on the edge of your glass (you want the "C" shape of the fork-spoon to wrap around the edge of the glass—not pointing the other way around). Once it's balanced, have your lab assistant (an adult) light both ends of the match.

You should be able to balance the fork-spoon-match system on the edge of your glass without holding it.

THE SCIENCE

 How in the world is this not falling?!

 Sometimes the center of mass of an object is actually outside the object itself!

 Huh?

 I know, right? This happens for irregularly shaped objects. So, for the fork-spoon system, the center of mass is slightly outside both the fork and the spoon. In fact, it's about a match-length away! We need the match to get its center of mass on top of the glass. The match is so light compared to the fork and the spoon that it doesn't really affect where the center of mass is.

 It just helps us balance the fork and the spoon?

 That's right! The match is the secret ingredient here that helps us balance this irregularly shaped object. →

149

TRY THIS →

Have an adult light both sides of the match. Notice how the system still doesn't fall? That's because the edges of the match weren't doing anything in terms of balance. The center of mass (the point that touches the glass) is still intact. So, the system stays balanced!

NOTES!

Twirling Dancer

How does a compass work?

ABOUT THE EXPERIMENT

Did you know that the Earth itself is a magnet? It's very big, but it's also very weak. A compass uses a little magnetized needle that can move really easily, which allows it to be moved by Earth's magnetic field. Because opposite charges attract, the southern pole of the needle will point to Earth's magnetic north pole. So, on a compass, the magnetic south pole of the needle will be painted red to show you which way is north.

With this experiment, we're going to create our own magnetic field using a battery and a magnet. I'll show you how to make a copper wire dance around a battery

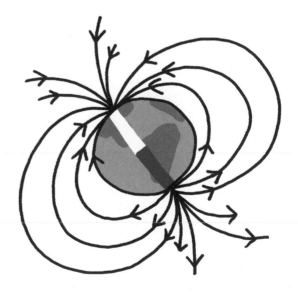

using something called the **Lorentz force**. This is one of the more advanced science experiments in this book, so make sure you have an adult to help you with this one. →

AA battery

4 disc magnets

Copper wire

Wire cutters

Paper, markers, and tape

(if you want to get creative)

INSTRUCTIONS

1. Stack your 4 disc magnets on top of each other.

2. Place the AA battery vertically on top of your four disc magnets.

3. Make your dancer with the copper wire. Get creative here! The trick is to make a copper figure that will sit on top of the battery and (ever so slightly) graze the sides of the magnet AND stay balanced while it spins. It's harder than it looks! I find it easiest to slowly remove the copper wire from its spool and create something that looks like a Christmas tree.

Safety note:
Make sure you have adult supervision here! Fingers can get pinched in between disc magnets, so watch your fingers. And make sure you never put magnets near any electronics or you could break them.

DID YOU KNOW?
↓

We see the northern lights in the NORTH because of Earth's magnetic field. The scientific name for the northern lights is the aurora borealis. An aurora occurs when the sun burps energetic particles toward the Earth. Those energetic particles run into Earth's magnetic field, which directs them (like cars on a highway) to the north and south part of our planet. When the particles get there, they interact with molecules in our atmosphere, which makes the beautiful colors you see in the aurora. The auroras that occur in the south are called the aurora australis. We just don't hear about these as often because there are fewer people who live in that region.

4. Make a little knob at the top of your copper figure to create a point of balance for the dancer. This knob will rest on the top of your battery.

5. Pull the bottom of the dancer down so that part of it ever so slightly touches the sides of your magnets. (You don't want it to be too tight around the magnets or else it won't have room to spin. You just want the copper figure to graze the magnets.)

6. If you want, draw a face and hands on some paper, cut them out, and tape them to the top and sides of your copper figure to create a "dancer."

You should see your copper figure spin on top of your battery. It may be hard for it to balance on your battery for longer than a few seconds. Make sure the bottom of your copper figure isn't too tight around the magnets—you want it to be able to spin easily around them.

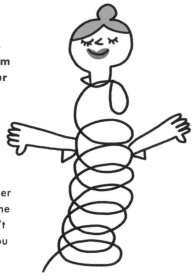

THE SCIENCE

 I finally got it to work! It was hard to balance it at first, but I got it!

 That's great! Science is all about trial and error, and you may have to change your design a little bit to get this one to work. It can be tricky because you need the copper wire to touch both the battery AND the magnet so that you close the circuit—this is what allows current to move through your wire. But you also don't want it to hold on to the magnet too tightly or else it won't be able to spin.

 Oh, that's why it was so hard! So, what made my dancer spin?

 Well, first the battery made current move through the copper wire. Copper is conductive, meaning electrons can easily move through it. It's like a highway for electrons! "Current" is just a fancy name for moving electrons.

 Gotcha—so my dancer has a bunch of electrons moving through them! What was the magnet for?

 The magnet is the secret sauce here. To create the Lorentz force, which will push the dancer to spin, we need a magnetic field. That's what the magnet is doing—it's creating a magnetic field.

 So, the current moved through the dancer, which was in the presence of a magnetic field created by the battery—and THAT created the Lorentz force that made my dancer spin?

 You got it!

DECK HALLS EXPER

THE
WITH
MENTS

The Incredible Growing Balloon

Colder Than Ice

Why do people put salt on the roads in the wintertime?

ABOUT THE EXPERIMENT

Salt lowers the **freezing point** of ice, so when we put salt on the road in wintertime, it helps melt the ice on the road, and it prevents more ice from forming. Let's see that with our own eyes, shall we? We're going to watch how salt affects the temperature of a glass of ice. It sounds simple enough, but you'll be surprised by how drastic the results are! The first time I did this one, it blew my mind!

Materials
↓

Food thermometer

2 glasses filled with ice

¼ cup (70 g) salt
(any kind will work)

INSTRUCTIONS

1. Use your food thermometer to measure the temperature in each of your glasses—they should be about 32 degrees Fahrenheit, or maybe slightly above because your glass is at room temperature and may have slightly warmed the ice. This is the freezing point of ice.

2. Now, put the salt in one of your glasses.

3. Wait 5 minutes.

After 5 minutes, your two glasses of ice will have melted a little bit, but one will look a little different than the other. Let's learn why!

THE SCIENCE

 What happened to each glass after 5 minutes?

 Ooh, the glass with salt has more water in it!

 That's right! Now, use your food thermometer to measure the temperature in each glass again.

 WHOA! Why did the glass with the salt get colder?

 Isn't that COOL? The cup with salt melted a little more AND got colder. You may even see frost on the outside. Two things are happening here: First, salt is lowering the freezing point of ice. That means it now needs to be colder to stay frozen. So, more ice is going to melt in the cup with salt. And second, it takes a lot of energy for ice to melt, and it takes that energy in the form of heat from its environment. It's essentially stealing heat from the environment so that it can have the energy to break apart and melt!

HEY! WE NEEDED THAT!

WHO TO KNOW
↓

Indigenous groups in the Arctic—including various Inuit groups in Alaska, Canada, and Greenland—have worked with traditional scientists to better understand how the climate has changed that region of the world over time. While traditional scientists can study the Arctic at a large scale with satellite imagery, Indigenous groups have a very detailed knowledge of sea ice in their region at a smaller scale.

Salty Ice

Why do potholes form in the road in the wintertime?

ABOUT THE EXPERIMENT

f you live in an area that experiences snowy winters, you've probably seen a pothole or two in the road. Ever wonder how those form? Well, even the most perfectly paved road will have some small gaps and cracks. When it rains, water can get into those gaps and if the ground is cold enough—like it is during the winter—that water will freeze. When water freezes, its volume gets bigger. That means that the water will push on those small cracks and expand them, little by little. This happens over and over again until a small crack turns into a pothole!

Potholes are nature's way of showing us that water can easily turn from a liquid to a solid to a liquid and back to a solid again. In this experiment, we'll see that transformation in action and use it to lift up a piece of ice with a string!

Materials
↓

Glass of
water

Ice cube

String
or yarn

Salt

(any kind will work)

INSTRUCTIONS

1. Fill your water glass to the top.
2. Add the ice cube to the glass.
3. Place the string on top of the ice cube.
4. Wait a couple of minutes, and then try to pick up the ice cube with the string (it doesn't work).
5. Now place the string back on top of the ice cube. Sprinkle salt all over the string and the ice cube.
6. Wait 2 to 3 minutes.
7. Try to pick up the ice cube with the string.

Your ice should be stuck to the string and you should be able to lift the ice up with the string.

THE SCIENCE

 Cool! How did salt help me pick up the ice cube?

 Well, salt lowers the freezing point of ice, so it helped melt the top layer of the ice. When ice melts, it steals heat from its environment, making it colder. This means that the water that the ice is sitting in got a little colder—so cold, in fact, that it FROZE!

 So it froze over the top of the string?

 Exactly, and that's what made it possible to pick up the ice cube with your string!

TRY THIS
↓

Try the same experiment with sugar. What happens? Well, sugar and salt BOTH lower the freezing point of ice, so it will work with either one. Salt is a little more effective at lowering the freezing point, though (which is why we use salt to prevent icy roads in the winter and not sugar).

The Incredible Growing Balloon

What happens to a balloon if you put it outside in the winter?

Lab Assistant (an Adult!) Required

ABOUT THE EXPERIMENT

I f you blow up a balloon, tie it off, and place it outside on a cold day, it will slowly deflate—even though it is completely closed! This is because the air inside the balloon gets colder and condenses (the air molecules slow down, so the air **pressure** gets lower, and overall, it takes up less space). If you bring it back inside, the air inside the balloon will warm up again— the air molecules inside will dance around more quickly, which increases the air pressure and makes the air take up MORE space! You can keep doing this again and again and again. The air pressure decreases when the balloon is cold and increases when the balloon is warm. This experiment will show you how to create the same effect indoors.

Materials

↓

Balloon

Plastic bottle

Mug of hot water
(you'll need adult supervision for this part)

Bowl of ice and water

INSTRUCTIONS

1. Make sure your plastic bottle is open (without the cap on top).

2. Place the balloon on top of your plastic bottle (no need to blow it up, just place it on top).

3. Have your lab assistant (an adult) place the bottle in a mug of hot water, and then wait 20 seconds.

4. Now, place your bottle in the bowl of ice and water.

5. Repeat as many times as you like.

As you place your bottle in the mug of hot water, your balloon should slightly fill with air and stand up straight. As you place your bottle in the bowl of ice water, the balloon will start to deflate and after about 30 seconds will completely deflate and fall down.

THE SCIENCE

 How did the balloon blow up?

 Well, your hot water heated up the air inside your bottle (and your balloon). When air gets hot, the molecules move around faster and more frequently, which increases the air pressure. This means that when the air gets hot, it wants to take up more space.

 So the higher air pressure made the balloon get bigger?

 That's right! The balloon is stretchy, so the increased air pressure was able to force the balloon to get bigger so that the air could take up more space. What did you notice when you put the bottle in the cold water?

 The balloon deflated again. Is that because the air got colder?

 Exactly! The air got colder, the molecules moved around more slowly, and the air pressure went down. All of this means that the air wanted to take up less space, so the balloon deflated!

TRY THIS

↓

Try blowing up a balloon and putting it in your freezer. How much smaller does it get? What happens when you bring it back into your warm room? How long does it take to shrink and grow? Those air molecules take a while to go to sleep (cool down) and wake up (warm up), don't they?

Electromagnetic Spectrum

The universe is filled with many different types of waves called electromagnetic radiation. These waves can be used to transfer energy and information. The Electromagnetic Spectrum shows you the full range of electromagnetic radiation, from large waves (like radio waves) to small waves (like gamma rays).

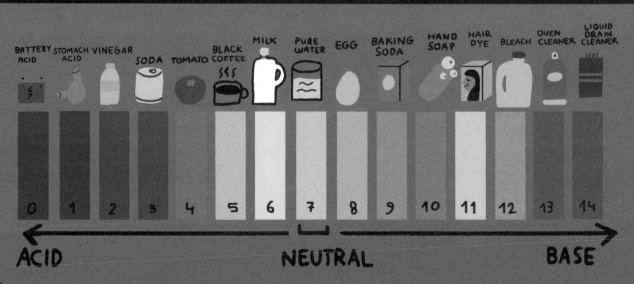

BATTERY ACID	STOMACH ACID	VINEGAR	SODA	TOMATO	BLACK COFFEE	MILK	PURE WATER	EGG	BAKING SODA	HAND SOAP	HAIR DYE	BLEACH	OVEN CLEANER	LIQUID DRAIN CLEANER
0	1	2	3	4	5	6	7	8	9	10	11	12	13	14

← ACID   NEUTRAL   BASE →

Charts

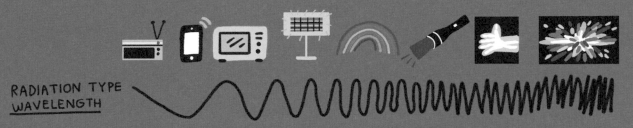

RADIATION TYPE
WAVELENGTH

RADIO

Radio waves are used by TV and cell phones

MICROWAVE

Microwaves are used by cell phones and microwaves

INFRARED

Things that are hot give off infrared light

VISIBLE

This is the light we see with our eyes

ULTRAVIOLET

The sun and black lights create ultraviolet waves

X-RAY

X-rays can be used to see through skin

GAMMA RAY

Gamma rays are created by really powerful events like supernovas

The pH Scale

The pH scale is a scale that tells you how acidic or basic a solution is. Something with a pH between 0 and 6 is acidic (with 0 being the most acidic), something with a pH of 7 is neutral, and something with a pH between 8 and 14 is basic (with 14 being the most basic).

Science Glossary

AURORA

An aurora is a beautiful display of lights in the upper atmosphere that occurs when really energetic particles travel from the Sun to Earth.

BACTERIA

Bacteria are microscopic single-celled organisms that you can only see under a microscope. There are more than 700 different types of bacteria living in your mouth right now! Yum!

BLACK HOLE

A black hole is a place in the universe where gravity is so intense that nothing can escape it—not even light!

CENTER OF MASS

The center of mass of an object is the average position of all the mass in the object.

CHEMICAL

A chemical is any substance that has matter. Sounds pretty general, doesn't it? Chemicals are not inherently good or bad and they make up all of the solids, liquids, and gases around us! Chemical compounds are found in everything from the food we eat to the crayons we draw with to the air we breathe!

CHEMICAL REACTION

A chemical reaction happens when one or more chemicals are combined together to create something new or different.

CIRCUIT

A circuit is the design of a path of two or more points where current can flow. An electrical circuit can include many different electronic components like batteries, lights, motors, resistors, and more! A closed circuit is a type of circuit where current can flow uninterrupted. An open circuit means there is a break in the circuit so that the current cannot flow.

COMBUSTION

Combustion is the process of burning something.

CONDUCTOR

A conductor is something that electricity can easily flow through.

CURRENT

Current is the movement of charged particles like electrons.

DENSITY

Density measures how much stuff is in a given volume. The equation for density is mass divided by volume (d = m/V). *d* is for density, *m* is for mass, and *V* is for volume. Something that is more dense than something else will have more mass in the same amount of volume.

EXOSKELETON

Exo means "outside," so an exoskeleton is a skeleton that exists outside of the body. Many animals like cockroaches, beetles, and scorpions have this type of protective structure on the outside of their bodies.

FLUORESCENT

A fluorescent material is something that shines really brightly when some form of light is directed onto it.

FREEZING POINT

The freezing point is the temperature at which a liquid will freeze into a solid (at atmospheric pressure).

FRICTION

Friction is the resistance that an object feels when moving across another object.

HEAT CAPACITY

The heat capacity of a substance refers to how much heat is required to raise that substance's temperature. Something with a high heat capacity requires a lot of heat to make that thing hotter.

HYDROPHILIC

Hydro refers to water and *philic* refers to having a fondness for something (it's the opposite of *phobic*, which is where *phobia* comes from). So this is a word that literally means "water loving."

LED

LED is an acronym that stands for light emitting diode. An LED emits light when electrical current flows through it.

LIPOPHILIC

Lipo refers to fat and *philic* refers to having a fondness for something (it's the opposite of *phobic*, which is where *phobia* comes from). So this is a word that literally means "fat loving."

LORENTZ FORCE

The Lorentz force is created when you have a current-carrying wire in the presence of a magnetic field.

MEMBRANE

A membrane is a thin layer or sheet that acts as a boundary. A semipermeable membrane is a type of membrane where certain molecules can slip right through the boundary to the other side.

METEORITE

A meteorite is a space rock that has fallen down to Earth.

NEUTRON STAR

A neutron star is created when a massive star runs out of fuel and collapses in on itself. It's a dense star that is made up of mostly neutrons.

NEWTON'S FIRST LAW OF MOTION

An object at rest will stay at rest unless a force is applied to that object. Also, an object in motion will remain in motion unless a force is applied to that object. This law is also known as the Law of Inertia.

NEWTON'S SECOND LAW OF MOTION

An object will accelerate (begin moving) when a force is applied to it. A force on an object with less mass will accelerate more than if you apply the same force to an object with more mass. This law can also be defined using the equation F = m*a. *F* is for force, *m* is for the mass of an object, and *a* is for acceleration.

NEWTON'S THIRD LAW OF MOTION

This law states that for every action there's an equal and opposite reaction.

NON-NEWTONIAN FLUID

This is a type of fluid that doesn't follow the normal fluid laws defined by Isaac Newton. It's an outlaw of the fluid world! Sometimes it acts like a solid and sometimes it acts like a liquid.

OSMOSIS

Osmosis is the process of molecules moving through a semipermeable membrane from a less concentrated solution to a more concentrated solution in order to try to equalize the concentrations on both sides of the membrane.

PHOSPHOR

This is a substance that shines visible light when it is exposed to ultraviolet light.

POLYMER

A polymer is a substance that is made up of long repeating chains of molecules.

PRESSURE

Pressure is the word we use to describe a force over a given area. In fact, the way you calculate pressure is with the formula $P = F/A$. *P* is pressure, *F* is force, and *A* is area. *Atmospheric pressure* is the name for pressure created by the weight (a type of force) of the atmosphere.

SOLVENT

A solvent is a substance that is able to dissolve certain things. The most common solvent is water because it is able to dissolve so many different things.

STATIC CHARGE

A static charge is a buildup of charge (negative or positive) on a surface.

SUPERNOVA

A supernova is a beautiful, powerful, and bright explosion of a star.

SURFACE TENSION

Surface tension is the tension on the surface of a liquid that is created due to the attraction between the molecules at the surface.

SURFACTANT

This is a chain-like molecule found in soaps and detergents. One end of a surfactant is hydrophilic, and the other end is lipophilic. It can break the bonds between water molecules and because of this it reduces surface tension in a liquid.

ULTRAVIOLET

This is the type of light that comes right after violet on the electromagnetic spectrum. It has a slightly shorter wavelength than violet light and we can't see this type of light with our human eyes.

VACUUM

A vacuum is a space without any matter, like air molecules.

VORTEX

A vortex is a mass of spinning fluid.

About the Author

Emily Calandrelli is the host and co-executive producer of Netflix's *Emily's Wonder Lab*. Each episode features Emily and a group of kid-scientists as they learn about STEAM through experiments and fun activities. Emily is also an executive producer and Emmy-nominated host of FOX's *Xploration Outer Space* and was a correspondent on Netflix's *Bill Nye Saves the World*.

Emily, who was named to *Adweek*'s "11 Celebrities and Influencers Raising the Bar for Creativity in 2017," is also an accomplished writer and speaker on the topics of space exploration, scientific literacy, and equality. Her chapter book series, the Ada Lace Adventures, centers around an eight-year-old girl with a knack for science, math, and solving mysteries with technology. The second book in the series, *Ada Lace Sees Red*, was included in the National Science Teachers Association's list of best STEM books for 2018. The third book, *Ada Lace, Take Me to Your Leader*, was part of the initiative from NASA, CASIS, and Story Time from Space, where the book was launched into space and read by an astronaut aboard the International Space Station to an audience of kids. Emily is also the author of the children's picture book *Reach for the Stars*, which came out in March 2022.

Emily has given talks about the importance of science literacy, the benefits of space exploration, and the challenges for women in STEM careers for clients like Google, Pixar, MIT, and Texas Instruments as well as dozens of K–12 schools across the nation. Her first two TEDx Talks, "I Don't Do Math" and "Space Exploration Is the Worst," have garnered over one million views on YouTube.

Prior to her work in science communication, Emily attended West Virginia University, where she received bachelor of science degrees in mechanical engineering and aerospace engineering, and MIT, where she received two masters of science degrees, one in aeronautics and astronautics and the other in technology and policy. Through her work, she wants to make science relatable, easy to understand, and more welcoming to historically excluded groups.

NOTES!